# the SNAP solution

## An Innovative Math Assessment Tool for Grades K–8

**KIRK Savage**  **JONATHAN Ferris**  **TOM Hierck**

Solution Tree | Press

Copyright © 2024 by Kirk Savage, Jonathan Ferris, and Tom Hierck

Materials appearing here are copyrighted. With one exception, all rights are reserved. Readers may reproduce only those pages marked "Reproducible." Otherwise, no part of this book may be reproduced or transmitted in any form or by any means (electronic, photocopying, recording, or otherwise) without prior written permission of the publisher.

555 North Morton Street
Bloomington, IN 47404
800.733.6786 (toll free) / 812.336.7700
FAX: 812.336.7790

email: info@SolutionTree.com
SolutionTree.com

Visit **go.SolutionTree.com/mathematics** to download the free reproducibles in this book.

Printed in the United States of America

Library of Congress Cataloging-in-Publication Data

Names: Savage, Kirk, author. | Ferris, Jonathan, author. | Hierck, Tom, 1960- author.
Title: The SNAP solution : an innovative math assessment tool for grades K-8 / Kirk Savage, Jonathan Ferris, Tom Hierck.
Description: Bloomington, IN : Solution Tree Press, 2024. | Includes bibliographical references and index.
Identifiers: LCCN 2024003793 (print) | LCCN 2024003794 (ebook) | ISBN 9781960574763 (paperback) | ISBN 9781960574770 (ebook)
Subjects: LCSH: Mathematics--Study and teaching (Elementary)--Evaluation. | Mathematics--Study and teaching (Middle school)--Evaluation. | Educational tests and measurements.
Classification: LCC QA135.6 .S38 2024  (print) | LCC QA135.6  (ebook) | DDC 372.7--dc23/eng/20240326
LC record available at https://lccn.loc.gov/2024003793
LC ebook record available at https://lccn.loc.gov/2024003794

---

**Solution Tree**

Jeffrey C. Jones, CEO
Edmund M. Ackerman, President

**Solution Tree Press**

*President and Publisher:* Douglas M. Rife
*Associate Publishers:* Todd Brakke and Kendra Slayton
*Editorial Director:* Laurel Hecker
*Art Director:* Rian Anderson
*Copy Chief:* Jessi Finn
*Senior Production Editor:* Sarah Foster
*Copy Editor:* Evie Madsen
*Proofreader:* Mark Hain
*Text and Cover Designer:* Julie Csizmadia
*Acquisitions Editors:* Carol Collins and Hilary Goff
*Assistant Acquisitions Editor:* Elijah Oates
*Content Development Specialist:* Amy Rubenstein
*Associate Editor:* Sarah Ludwig
*Editorial Assistant:* Anne Marie Watkins

**SNAP!** *That's the sound of a person's thumb and middle finger rubbing together quickly, branches succumbing to the weight of ice on a frigid day, or burning wood releasing trapped moisture deep within a crevice. SNAP is a playful expression people use when surprised or when something previously unknown makes sense. It's the sound a camera makes when taking a snapshot, and it also stands for* student numeracy assessment and practice (SNAP). *A team of educators in Chilliwack, British Columbia, Canada, created the SNAP tool for numeracy assessment, which aims at helping teachers become more confident and flexible in assessing students' number sense. In other words, making a mathematics assessment a SNAP!*

# Acknowledgments

This book was written with the support of the Chilliwack School District. We thank and acknowledge the many dedicated educators in the district for developing the SNAP (including the creation and development of the templates, and piloting and implementing the SNAP to foster and support student learning).

We also want to thank the following consultants:

- Anna Webb is an original member of the numeracy assessment working group in Chilliwack, British Columbia, Canada, which created the SNAP. She has extensive experience in the classroom as a primary and intermediate teacher. Anna is a district curriculum mathematics support teacher, helping other teachers in Chilliwack.
- Christian Lodders is a district curriculum mathematics support teacher, helping other teachers in Chilliwack. He has extensive teaching experience as an upper-intermediate teacher.

The coauthors also want to thank the following contributors to this book:

- Andy Fast, grade 4 teacher at Vedder Elementary
- Paul Wojcik, grade 4 teacher at Vedder Elementary
- Heather McCague, kindergarten teacher at Strathcona Elementary

- Kara Rideout, grade 1 teacher at Strathcona Elementary
- Nancy Ferris, kindergarten teacher at Promontory Heights Elementary

The Chilliwack School District SNAP design team consists of the following people:

- Christine Blessin
- Jonathan Ferris
- Kathy Isaac
- Anna Webb
- Shannon McCann
- Tammy McKinley
- Justin Moore
- Kirk Savage
- Paul Wojcik

Solution Tree Press would like to thank the following reviewers:

Kelly Hilliard
GATE Mathematics
Instructor NBCT
Darrell C. Swope Middle School
Reno, Nevada

Erin Kruckenberg
Fifth-Grade Teacher
Jefferson Elementary
Harvard, Illinois

Paula Mathews
STEM Instructional Coach
Dripping Springs ISD
Dripping Springs, Texas

Janet Nuzzie
District Intervention Specialist
Pasadena ISD
Pasadena, Texas

---

Visit **go.SolutionTree.com/mathematics** to download the free reproducibles in this book

# Table of Contents

**ABOUT THE AUTHORS** .................................... xi

**INTRODUCTION** .......................................... 1
Moving Away From Traditional Assessment Methods .......... 3
Moving Toward the SNAP ................................... 5
About This Book .......................................... 7

**CHAPTER 1**

**Exploring the SNAP** .................................... 9
Why the SNAP Works ....................................... 10
Five Strands of Mathematical Proficiency ................. 19
Unique Features of the SNAP .............................. 19
Classroom Applications ................................... 23
Chapter Summary .......................................... 27

**CHAPTER 2**

**Looking at Number Sense Foundations** ................... 29
A Theoretical Perspective on Number Sense ................ 30
Number Sense in the Early Years: Kindergarten ............ 31

Number Sense in Grades 1–5 .................................. 33
Number Sense in Grades 6–8 .................................. 36
Classrooms as Safe Spaces .................................... 40
Chapter Summary ............................................ 40

**CHAPTER 3**

## Understanding the Five Strands of Mathematical Proficiency ........................ 41

Conceptual Understanding .................................... 43
Procedural Fluency ........................................... 46
Strategic Competence ........................................ 51
Adaptive Reasoning .......................................... 55
Productive Disposition ....................................... 57
Chapter Summary ............................................ 59

**CHAPTER 4**

## Implementing the SNAP With Beginning Mathematicians ...................... 61

Implementing the SNAP in Kindergarten and Grade 1 ........ 61
Using the SNAP in Kindergarten .............................. 63
Using the SNAP in Grade 1 ................................... 66
Chapter Summary ............................................ 70

**CHAPTER 5**

## Exploring Rubrics, Assessment, and Competency-Based Learning .................... 71

Understanding the SNAP Rubric ............................... 71
Class Profile Example of a Mixed-Number Task .............. 73
Class Profile Example of a Whole-Number Task .............. 83
Chapter Summary ............................................ 94

**CHAPTER 6**

## Understanding How the SNAP Supports Response to Intervention ...................... 95

Supporting RTI .............................................. 96
Supporting Tier 1 ............................................ 99
Supporting Tier 2 ........................................... 105
Supporting Tier 3 ........................................... 108
Chapter Summary ........................................... 108

## CHAPTER 7

### Looking at School and District SNAP Implementation ... 109

Implementation at the School or District Level ... 110
Data Analysis and Monitoring ... 111
Cut Scores ... 112
Transparency and Accountability ... 112
Training ... 113
School Leadership ... 113
Chapter Summary ... 114

## EPILOGUE

### Final Word ... 115

## APPENDIX A

### SNAP Templates, Rubrics, and Classroom Profiles ... 117

*Kindergarten Number Sense SNAP* ... 118
*Grade 1 Number Sense SNAP* ... 119
*SNAP for 0–100* ... 120
*SNAP for 0–1,000* ... 121
*SNAP for 0–10,000* ... 122
*SNAP for 0–100,000* ... 123
*Open-Ended SNAP* ... 124
*Blank SNAP* ... 125
*SNAP Number Sense Rubric* ... 126
*SNAP Number Sense Class Profile* ... 127

## APPENDIX B

### Resources to Support Number Sense ... 129

Books ... 129
Websites ... 132

### REFERENCES AND RESOURCES ... 135

### INDEX ... 145

# About the Authors

 **KIRK SAVAGE, EdD,** is an assistant superintendent for Chilliwack School District in British Columbia, Canada, where he has been working as a senior leadership team member since 2012. Kirk has also led in the role of elementary and middle school principal and has taught at all levels, from elementary to secondary. He was a part-time secondee to the British Columbia Ministry of Education for ten years, developing provincial numeracy and literacy assessments. In 2017, he was a ministry outreach ambassador, assisting districts with implementing British Columbia's competency-based curriculum.

Kirk's passions are curriculum and instruction, particularly the connection between instructional design and authentic assessment. He has coauthored two books, *The ANIE: Assessment of Numeracy in Education* and *P.L.A.N. for Better Learning*, which expand on his passion for using practical classroom examples that impact teaching efficacy, student engagement, and results.

Kirk is a system leader who always strives for improvement while leading change. Whether it is strategic-plan development, middle school reconfiguration, or response to intervention (RTI) implementation, he has the experience and expertise to lead improved student achievement and renewed culture in multiple contexts.

Kirk lives with his wife and two children in Chilliwack. He is a retired professional and national team Ultimate Frisbee athlete and a 2017 inductee to the Canadian Ultimate Hall of Fame.

**JONATHAN FERRIS, MA,** has been an educator since 1990. His passion for student learning has driven many innovative contributions in his school community, including cochairing the SNAP and ACT assessments. Through his work as an education assistant, classroom teacher, district helping teacher, principal, and educational consultant, Jonathan amassed a wealth of understanding, skill, and knowledge. He is a program coordinator at the University of the Fraser Valley (in Abbotsford, British Columbia, Canada), where he oversees practicum placements for teacher candidates, collaborates with faculty mentors, and teaches courses on classroom management, professionalism, ethics, and teacher presence.

Jonathan is a member of a collaborative learning committee (composed of members from the University of the Fraser Valley and Chilliwack School District) that facilitates joint learning opportunities between the two learning organizations. When he was a principal, Jonathan's school was featured in *The Globe and Mail* newspaper for innovation regarding awards assemblies, and on Global News Television for student engagement during the COVID-19 shutdown. Jonathan wrote the foreword to *P.L.A.N. for Better Learning* by Kevin Bird and Kirk Savage, and coauthored chapters in *Online Learning and Teaching From Kindergarten to Graduate School* and the forthcoming *Pedagogies of Practicum: Post-Pandemic Reflections on Innovations in Practice Teaching*.

Jonathan earned a bachelor's degree in education from the University of British Columbia and a master of arts degree in leadership in curriculum and instruction from San Diego State University.

**TOM HIERCK** has been an educator since 1983 in a career that has spanned all grade levels and many roles in public education. His experiences as a teacher, an administrator, a district leader, a department of education project leader, and an executive director have provided a unique context for his education philosophy.

Tom is a compelling presenter, infusing his message of hope with strategies culled from the real world. He understands that educators face unprecedented challenges and knows which strategies will best serve learning communities. Tom has presented to schools and districts across North America with a message of celebration for educators seeking to make a difference in the lives of students. His dynamic presentations explore the importance of positive learning environments and the role of assessment in improving student learning. His belief that every student is a success story waiting to be told has led him to work with teachers and administrators to create positive school cultures and build effective relationships that facilitate learning for all students.

His works include *Trauma-Sensitive Instruction: Creating a Safe and Predictable Classroom Environment* and *Trauma-Sensitive Leadership: Creating a Safe and Predictable School Environment* (both with John F. Eller), and *The Road to Success With MTSS: A Ten-Step Process for Schools* (with Chris Weber).

To book Kirk Savage, Jonathan Ferris, or Tom Hierck for professional development, contact pd@SolutionTree.com.

# Introduction

When you mention *math* in a casual conversation, people may furrow their brows, stifle a yawn, or politely ask to change the subject. Discussing mathematics fits alongside taboos (such as a deliberation about religion or politics) most people do not talk about. Ask elementary students what their favorite subject is in school, and they will likely reply, "PE" (physical education). Ask them to name their least favorite subject, and they will invariably answer, "Math." "Math, more than any other subject, engenders anxiety and avoidance in students" (Shore, 2015). Ask elementary teachers which subject they most enjoy teaching, and they will usually reply, "Language arts." Ask them about their least favorite subject, and many will likely answer, "Math." Therefore, changing the students' response often begins with changing the teachers' attitudes because, as doctoral student Barbara L. Johnson (2018) suggests, "Although early childhood and elementary educators teach mathematics to children, many of them possess negative attitudes and beliefs about mathematics such as mathematics anxiety and low mathematics self-concept" (p. 1).

It would not be a stretch to say many people have "mathematics wounds" or their interactions with this subject have left them with "scars" from a myriad of poor experiences at an early age. Wouldn't it be extraordinary if there were a way for students to access mathematics through an assessment that is accessible, full of wonder, and provides a salve for the mathematics wounded?

Why is it that in some classrooms, students groan or engage in avoidance tactics when a teacher announces it is time for a mathematics lesson, when these same students seem energized and enthusiastic in other classrooms? Why do people often sympathize when someone says they are "not good at mathematics"? How many teachers have heard the lament, "I'm not good at mathematics, my parents weren't good at mathematics, and neither were my grandparents. I come from a long line of mathematics-defective people." People simply accept that "not being good at mathematics" is a real and serious issue that can have profound impacts on a person's ability to manage their adult responsibilities (for example, money management, job skills, interpreting data, and so on). Moreover, Youngstown State University researchers Molly Jameson and Brooke Fusco (2014) reveal that "despite the importance of mathematical understanding, many adults dislike and avoid mathematics, even those who are competent in mathematics. The reason for this dislike and avoidance is likely a combination of mathematics anxiety and low confidence" (p. 4). For many, these negative experiences remain throughout adulthood. Jo Boaler (2012), a leading mathematics education expert, reveals, "Math anxiety affects about 50 percent of the U.S. population and more women than men" (p. 1). So why do some individuals appreciate the elegance of numbers and the intrigue of mathematical reasoning, while others lose their sense of wonder for mathematics?

Students' attitudes toward mathematics shift as they progress through each grade. In kindergarten, students are full of mathematical wonder and inquisitiveness. They arrive at school ready to explore their environment, including the fascinating world of numbers. Sadly, and far too often, the playful aspect of manipulating objects to demonstrate quantities takes a back seat to filling out worksheets with endless practice, and as children reach grade 2 or 3, a great deal of curiosity with mathematical concepts has been lost (Fisher, 2005; Fosnot & Dolk, 2001). Mathematics appears to have been reduced to a series of algorithms to memorize, where much work is done with little thought (Burns, 2007; Fosnot & Dolk, 2001; Seely, 2009). We wrote this book to help teachers ensure a positive disposition toward mathematics continues throughout school and beyond. We believe the fun of playing with numbers and toying with mathematical concepts are things educators can foster—as long as they don't allow mathematics to become rote with tasks that focus on one way of solving equations, memorizing facts without understanding them, taking timed tests, and following a rigid set of rules (Mensah, Okyere, & Kuranchie, 2013). Educators want to light the fire of curiosity in students so they look at the world through a lens of optimism and possibility.

As powerful as a positive disposition in mathematics is, the opposite is also true. Those mathematics-wounded adults can impact future generations. If they become teachers, their negative bias toward mathematics can perpetuate the numeracy aversion cycle among their students in the same way that a positive mindset does just the opposite, and "if people who are anxious about math are charged with teaching others mathematics—as is often the case for elementary schoolteachers—teachers' anxieties could have consequences for their students' math achievement" (Beilock, Gunderson, Ramirez, & Levine, 2010). Educator, activist, and author Parker J. Palmer (2017) states that "good teaching cannot be reduced to technique; good teaching comes from the identity and integrity of the teacher" (p. 184). The enthusiasm and energy educators bring to their teaching can be contagious, especially when they have a genuine interest in the topic. Mathematics, with its precise logic and intricate patterns, gives educators unparalleled opportunities to amaze and inspire students. In essence, mathematics goes far beyond mere formula memorization; it's a gateway to the joy of working with numbers and applying them in everyday life.

## Moving Away From Traditional Assessment Methods

How educators structure lessons can have a profound effect on how well students learn. Three common approaches for lesson design are (1) transmission, (2) discovery, and (3) connectionist (Askew, Rhodes, Brown, Wiliam, & Johnson, 1997).

When a teacher presents a lesson using a *transmission* approach, the teacher asks students to open their textbooks while the teacher explains the lesson and writes questions on a whiteboard. Students then practice on their own. Next, the teacher provides time for questions and then assigns a series of problems for the students to undertake in relative silence—often with the unanswered questions left for students to complete as homework.

Using the *discovery* approach, the teacher may put out a set of manipulatives for students to work with to deduce patterns, themes, or concepts. Teachers encourage students in small groups to work together to explore relationships and mathematical ideas.

Brain research reveals the *connectionist* approach allows for deeper learning, and that students need to connect with their learning, process their learning in scaffolded ways, transform their thinking by engaging in high-inference tasks, and then reflect on their learning (Close, 2001). Using this approach, teachers connect students to the lesson by posing a question or treating an algorithm as a puzzle to solve. The teacher assists the students in processing the new information (through scaffolded tasks), and then students work in small groups to answer the question or solve the equation. Discourse among peers is welcome, and the teacher rotates between groups to monitor and check students' thinking, often stopping to address the whole class when a group makes a new connection. After the students master the task, the teacher challenges them to transform their thinking by creating real-world scenarios where they can apply their new connections. At the conclusion of the lesson, students reflect on their learning by asking questions such as, "Did I meet my thinking goal?" "What new connections did I make?" and "What part of the lesson was easy or challenging for me?"

Looking back at schooling in traditional classrooms in the 1970s and 1980s, mathematics assessment was a source of dread for some students. As assessment expert Richard J. Stiggins (2002) states, "With respect to the use of assessment to motivate, we all grew up in classrooms in which our teachers believed that the way to maximize learning was to maximize anxiety, and assessment has always been the great intimidator" (p. 760). Pop quizzes, timed drills, the infamous *mad minutes* (or speed drills), or unit tests conjure uneasy memories long after students leave school. Word problems, in particular, held the unique title of producing very high levels of anxiety and confusion for students (Khoshaim, 2020). Traditional mathematics instruction often involves students taking a weekly quiz (similar to a spelling test on Fridays), and then a chapter test at the conclusion of each unit. Typically, teachers average scores, with the concluding test carrying the most weight. This approach to instruction perpetuates a cycle of futility and pressure—effectively extinguishing the joy of learning and replacing it with anxiety and apathy. Knowing how educators assessed mathematics in the past, educators today are eagerly seeking new ways of assessing mathematics in ways that both support student learning and enhance engagement.

The primary aim of this book is to empower educators—especially those seeking effective teaching and assessment methods—to uplift students by fostering both wonder and fascination in the learning process. As coauthors, our desire is to both promote mathematical understanding in number sense and prevent students from feeling mathematics "wounded."

Fortunately, traditional practices are becoming less prevalent as teacher pedagogy changes in reaction to competency-based curricula. Student-centered educators see assessment as ongoing, and they incorporate daily observations, journals, and real-life applications to assess student learning and assist in planning further instruction. Researchers identify two main types of assessment: traditional, with a reliance on rote learning, and competency-based, which promotes *as*, *of*, and *for* learning. Thought leaders, such as Dylan Wiliam (2018), support the latter. Educators need to ensure feedback causes a *cognitive* rather than an *emotional* reaction; in other words, feedback should cause *thinking*. Wiliam (2018) explains:

> To be effective, feedback needs to direct attention to what's next, rather than focusing on how well or poorly the student did on the work, and this rarely happens in the typical classroom.... If we are to harness the power of feedback to increase student learning, then we need to ensure that feedback causes a cognitive rather than emotional reaction—in other words, *feedback should cause thinking* by creating desirable difficulties. (pp. 143, 152)

Student learning has predominantly become competency and performance based regardless of jurisdiction or school district. American, Canadian, Australian, and New Zealand education have all moved to such models for their curricula. Whether the curriculum is based on the Common Core State Standards (United States), Core Competencies (British Columbia), Achievement Standards (Australia), or Key Competencies (New Zealand), school systems have moved toward a competency- and performance-based curriculum. However, easy-to-use competency-based assessments are missing from these frameworks. This is where the student numeracy assessment and practice (SNAP) comes in (Chilliwack School District, 2016)!

## Moving Toward the SNAP

What if a competency-based numeracy assessment asked students to work with numbers in flexible ways, invoked curiosity, and encouraged students to create personalized word problems connected to their lives? What if educators could recapture the wonder of mathematics for students and teachers, and fuel inspiration by creating a visually appealing assessment that allows for student creativity and elicits deeper thinking? What if this assessment was independent of students' reading ability as a component to demonstrate understanding? A team of educators from the Chilliwack School District (2016) in the Canadian province British Columbia posed these guiding questions as a challenge to create an assessment to address fluency, number work, and comprehension in a way that educators can use *for*, *as*, and *of* learning, and be visually appealing, engaging, and easy for a teacher to administer. Led by educators and administrators Kirk Savage and Jonathan Ferris, the team included a passionate group of elementary and middle school teachers who met bimonthly using a think tank structure, which allowed members to consider all opinions. The team developed the SNAP using the guiding principle, *If students in each grade master the foundational learning standards for number sense, they are then set up for success in later grades.* Since teaching number sense as a foundational concept in schools is crucial, can educators evaluate number sense in a way that is replicable, reliable, efficient, and effective? They can with the SNAP!

The *SNAP* is a one-page assessment tool that offers a practical approach to evaluating students' number sense (see figure I.1, page 6). This simple and reliable approach benefits classroom teachers because they can quickly implement it in their daily teaching practices. Moreover, this approach is less correlated with reading ability than traditional assessments. For instance, typical academic *competency assessments* (or assessments of thinking) require a great deal of reading to create the context for the task. Furthermore, traditional mathematics word problems tend to be *closed assessments*, meaning there is generally a right or wrong way to solve the problem. The SNAP is the opposite; this assessment does not require much reading, and because it is an *open assessment*, there is no one right way for students to show their learning. Students have unlimited freedom to communicate their thinking using the SNAP. Additionally, educators can use this assessment *for*, *as*, and *of* learning—meaning teachers can use the SNAP diagnostically, formatively, summatively, and as a teaching experience (Chilliwack School District, 2016). This book takes you through the SNAP step by step, from what it is to how a school or district can implement it.

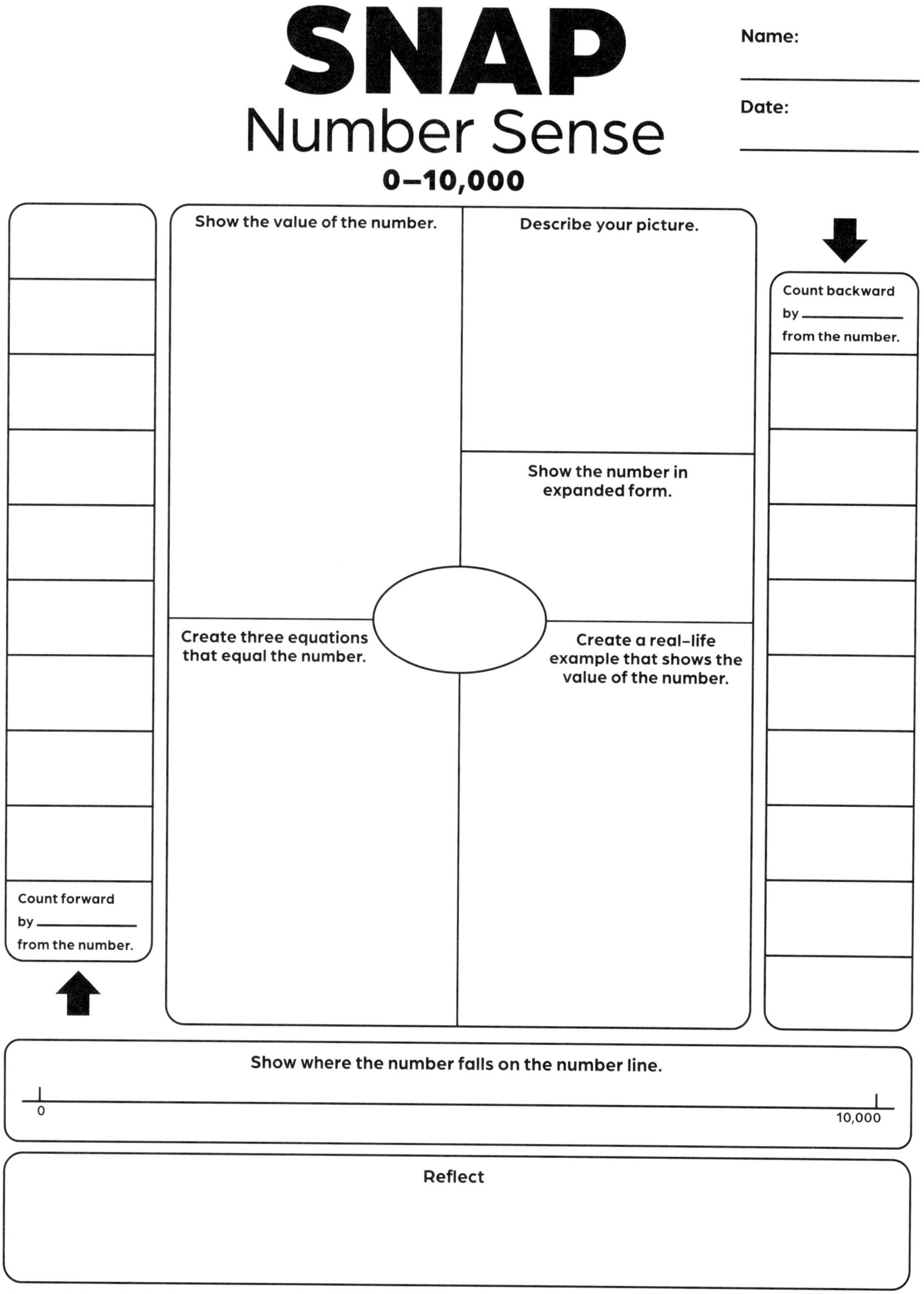

# SNAP
## Number Sense
### 0–10,000

Name: _____

Date: _____

| Show the value of the number. | Describe your picture. |

Count backward by _____ from the number.

Show the number in expanded form.

Create three equations that equal the number.

Create a real-life example that shows the value of the number.

Count forward by _____ from the number.

Show where the number falls on the number line.

0 ———————————————————— 10,000

Reflect

*Source: Chilliwack School District, 2016. Used with permission.*

**FIGURE I.1:** The student numeracy assessment and practice (SNAP).

*Visit **go.SolutionTree.com/mathematics** for a free reproducible version of this figure.*

## About This Book

In chapter 1, we explain what the *SNAP* is and how a small group of passionate educators seeking to create a new district assessment for number sense developed it. We also reveal how the SNAP aligns with the five strands of mathematical proficiency in the National Research Council's (2001) *Adding It Up: Helping Children Learn Mathematics*, and how the ANIE (Assessment of Numeracy in Education) inspired the SNAP (Bird & Savage, 2014). We share a completed student sample of the SNAP and discuss how teachers can use it in the classroom. Finally, we will highlight the unique characteristics of this assessment tool, which sets it apart from others.

In chapter 2, we discuss the foundations of number sense and include different grade-level practitioners' perspectives on the importance of number sense in the classroom. This chapter also includes the research underpinnings for the SNAP.

In chapter 3, we present the SNAP from the perspective of the five strands of mathematical proficiency. Readers will see how each skill from the SNAP aligns with a particular strand, including how to address each skill from the perspectives of early primary, early intermediate, and late intermediate educators.

Chapter 4 provides an overview of the SNAP from a kindergarten and grade 1 perspective, and details how primary teachers can adapt the SNAP to their grade levels. We also explain ways to address each skill in each section of the kindergarten and grade 1 SNAP.

Chapters 5, 6, and 7 address the big picture of the SNAP. In chapter 5, we explore how the SNAP aligns with the response to intervention (RTI) framework. In chapter 6, we delve further in by exploring rubrics, assessment, and competency-based learning. Finally, in chapter 7, we demonstrate how a school district can move forward with the SNAP implementation.

The book also gathers resources in two appendices. Appendix A collects blank versions of the SNAP. Appendix B includes resources that support number sense.

The SNAP's flexibility and alignment with many teaching approaches make it at home in any classroom. It shines when completed individually, or in small groups, a whole class, or as a primary strategy in individual education plans (IEPs).

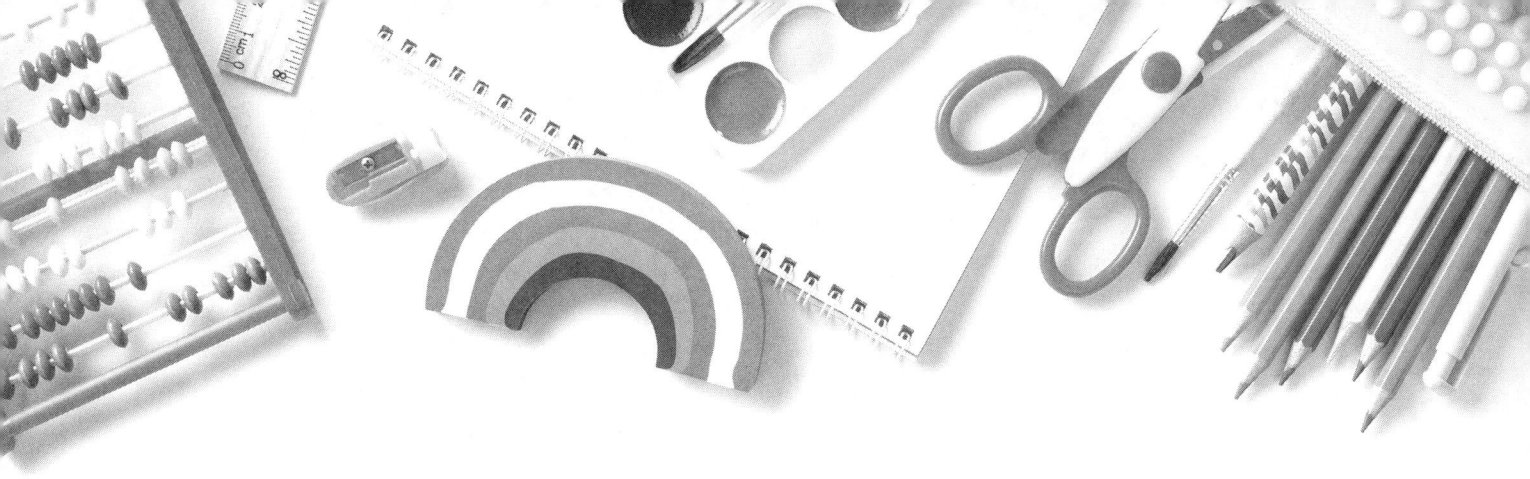

# CHAPTER 1

# Exploring the SNAP

The *SNAP* (*student numeracy assessment and practice*) is a one-page assessment tool designed to deepen students' understanding of number sense. *Number sense* is:

> A person's general understanding of numbers and operations along with the ability and inclination to use this understanding in flexible ways to make mathematical judgements and to develop useful strategies for handling numbers and operations. It reflects an inclination and an ability to use numbers and quantitative methods as a means of communicating, processing and interpreting information. It results in an expectation that numbers are useful and that mathematics has a certain regularity.
> (McIntosh, Reys, & Reys, 1992, p. 3)

Number sense is foundational to mathematical understanding and integral in developing mathematics fluency. Leading mathematics expert Jo Boaler (2015) suggests looking at fluency beyond purely focusing on speed and memorization, with the understanding that building meaning is most important:

> Teachers should help students develop math facts, not by emphasizing facts for the sake of facts or using "timed tests" but by encouraging students to use, work with and explore numbers. As students work on meaningful number activities they will commit math facts to heart at

> the same time as understanding numbers and math. They will enjoy and learn important mathematics rather than memorize, dread and fear mathematics. (p. 6)

With the SNAP, educators can use an assessment that is not only unbound by time but that also encourages students to be flexible with numbers by promoting different pathways of learning and invoking curiosity. We discuss number sense in greater detail in chapter 2 (page 29).

The SNAP's game-like appearance invokes a sense of wonder and invites students to take part in a series of challenges that promote flexible thinking in deep and connected ways. With a nonlinear design, the SNAP is much different from traditional worksheets. When using worksheets, some students feel secure in their knowledge, and others feel bored. In contrast, the SNAP provides *multiple access points* for students to display a wide variety of thinking skills. Being a one-page assessment (and with a game-like appearance), the SNAP does not overwhelm students with long lists of questions, and its open-ended nature provides enough challenge to deeply engage students' thinking. This chapter discusses why the SNAP works, the five strands of mathematical proficiency, the unique features of the SNAP, what research shows about assessment, and classroom applications for the SNAP.

## Why the SNAP Works

When students first receive the SNAP, they will notice it looks like a board game. To introduce it, teachers explain that the SNAP is a way for students to demonstrate their understanding of numbers. The object (or the game) of the SNAP is to use a number (in the oval) and show the different ways a series of challenges or tasks explains the number. These tasks include (in no particular order): draw to represent the value of the number, write to describe your picture, write the number in expanded form, create three equations that equal the number, write a real-life example that shows the value of the number, count forward and backward by the number, show where the number belongs on a number line, and finally, reflect on your thinking.

Ideally, teachers use the SNAP to assess and teach a wide variety of mathematics outcomes. Knowing that the SNAP has several functions, teachers can place other concepts in the center oval. For example, when working with geometry, a teacher could place a scalene triangle in the center and have students discuss the measurements of each angle. The ceiling is very high, allowing for greater creativity.

Here, we begin with number sense. Teachers place a number (such as 234) in the center of the SNAP, and then students perform a series of tasks reminiscent of moving through a board game. Each set of tasks is grounded in the curriculum, and each task focuses on different aspects of number sense (see figure 1.1).

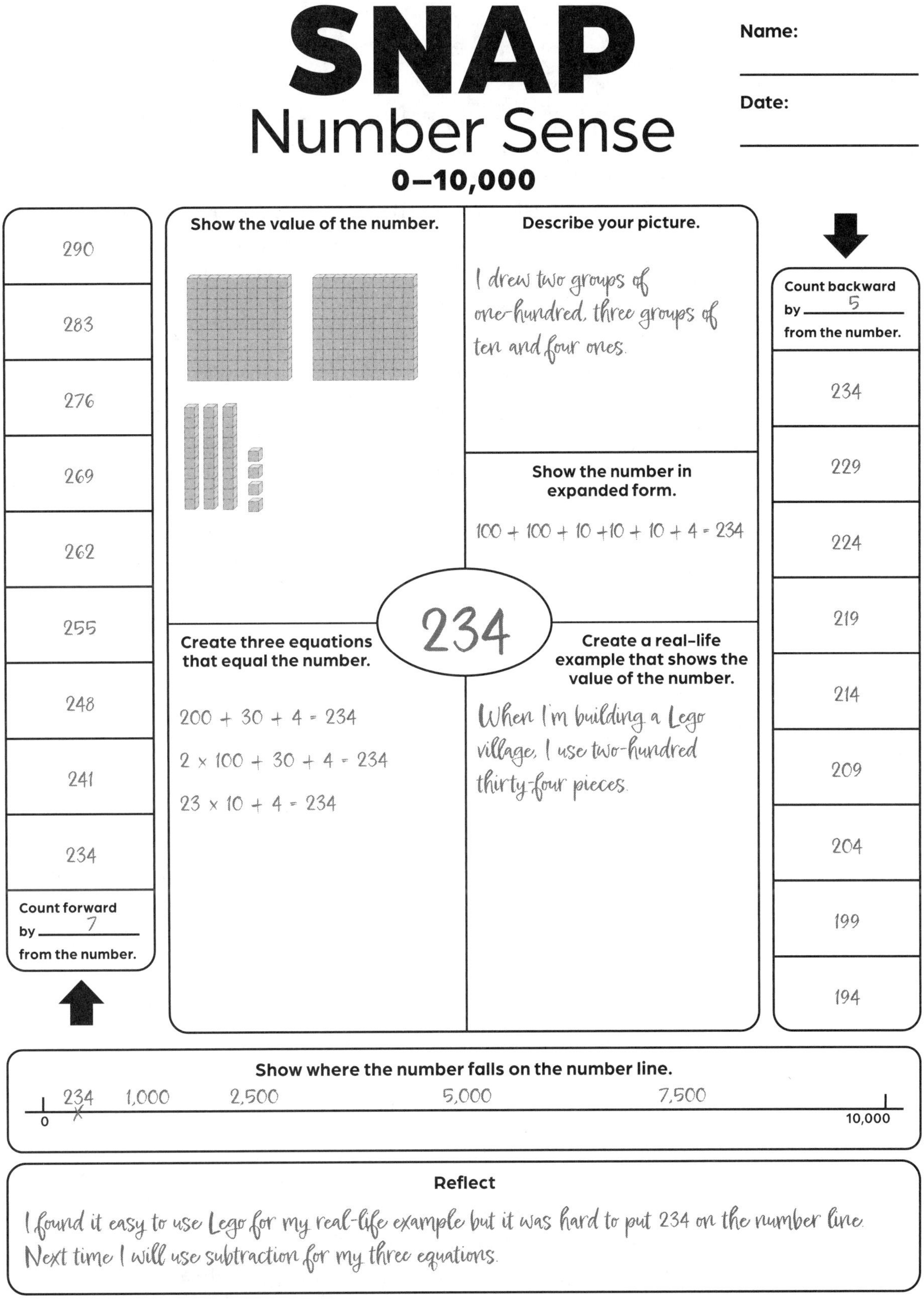

# SNAP
## Number Sense
### 0–10,000

Name: _____

Date: _____

| 290 |
| 283 |
| 276 |
| 269 |
| 262 |
| 255 |
| 248 |
| 241 |
| 234 |

**Count forward by** __7__ **from the number.**

**Show the value of the number.**

**Describe your picture.**
I drew two groups of one-hundred, three groups of ten and four ones.

**Show the number in expanded form.**
$100 + 100 + 10 + 10 + 10 + 4 = 234$

**234**

**Create three equations that equal the number.**
$200 + 30 + 4 = 234$
$2 \times 100 + 30 + 4 = 234$
$23 \times 10 + 4 = 234$

**Create a real-life example that shows the value of the number.**
When I'm building a Lego village, I use two-hundred thirty-four pieces.

**Count backward by** __5__ **from the number.**

| 234 |
| 229 |
| 224 |
| 219 |
| 214 |
| 209 |
| 204 |
| 199 |
| 194 |

**Show where the number falls on the number line.**
0 — 234(X) — 1,000 — 2,500 — 5,000 — 7,500 — 10,000

**Reflect**
I found it easy to use Lego for my real-life example but it was hard to put 234 on the number line. Next time I will use subtraction for my three equations.

*Source: Chilliwack School District (2016). Used with permission.*
**FIGURE 1.1:** The SNAP example for grade 4.

Figure 1.1 demonstrates a completed SNAP. We recommend teachers begin with just one task and then explicitly teach each area. Having students complete every task on the SNAP in one sitting depends on the grade level and time of year. From this example in figure 1.1, it is evident the student has a clear grasp of each task. Educators can glean a great deal of information through analyzing each completed SNAP. They could encourage students who completed figure 1.1 to stretch their thinking in areas such as writing a real-life example and reflection. The skills the team chose to include are grounded in the curriculum; we'll explain those skills in the upcoming sections.

## A FOCUS ON NUMBER SENSE SKILLS

One benefit of using the SNAP is that it allows students to hone these skills: draw to represent the value of the number, write to describe your picture, write the number in expanded form, create three equations that equal the number, write a real-life example the shows the value of the number, count forward and backward by the number, show where the number belongs on a number line, and reflect on your thinking to reinforce each aspect of number sense. Students work with the number in the oval as they proceed through the SNAP. Each skill can stand alone, but teachers must provide scaffolded teaching to ensure students become familiar with each area. There is no starting point on the SNAP. For example, students may feel more comfortable beginning with the number line instead of drawing the number. Therefore, when teachers first introduce the SNAP to students, it is critical to be flexible with the starting lessons, allowing for multiple entry points into learning the assessment.

*Number sense* contains a series of subskills for the students to master. In their review of educational research on number sense, coauthors Ian Whitacre, Bonnie Henning, and Şebnem Ataba; (2020) identify the following six primary skills relating to number sense.

1. *Number recognition* relates to being able to identify a number with words or symbols (Andrews & Sayers, 2015; Baroody, Eiland, & Thompson, 2009; Baroody, Eiland, Purpura, & Reid, 2012).

2. *Counting* is defined as rote counting (including the ability to count forward and backward from a number). *Number patterns* are associated with the ability to identify a missing number in a sequence (Andrews & Sayers, 2015; Jordan, Glutting, Dyson, Hassinger-Das, & Irwin, 2012; Jordan, Glutting, & Ramineni, 2010; Jordan, Glutting, Ramineni, & Watkins, 2010; Jordan, Kaplan, Locuniak, & Ramineni, 2007; Jordan, Kaplan, Oláh, & Locuniak, 2006).

3. *Number patterns recognition* is defined as a list of numbers that follow a defined sequence. Naming numbers, patterning numbers, identifying missing numbers, and comparing magnitudes all correlate with positive arithmetical learning by the end of the first year of schooling (Clarke & Shinn, 2004).

4. *Number comparison* is being able to define the difference between a set of numbers (Andrews & Sayers, 2015; Howell & Kemp, 2005, 2009, 2010).

5. *Number operations* refers to the ability to perform calculations (Andrews & Sayers, 2015; Baroody et al., 2009, 2012).

6. *Estimation* refers to being able to infer the quantity of a number, including the ability to identify the location of a number using a number line (Andrews & Sayers, 2015; Ivrendi, 2011; Jordan et al., 2012; Jordan, Glutting, & Ramineni, 2010, Jordan, Glutting, Ramineri, & Watkins, 2010; Jordan et al., 2006).

In addition, there are benefits to writing in mathematics. For example, researchers and coauthors Margaret Mills and Patricia Stevens (1998) had their students write word problems based on their interests and hobbies, then share them with their classmates. Mills and Stevens (1998) find this intervention had positive effects on student attitudes about short-answer mathematical writing. Furthermore, writing in mathematics can also help students consolidate their thinking because it requires them to reflect on their work and clarify their thoughts about the ideas (National Council of Teachers of Mathematics [NCTM], 2000). In addition, the NCTM (2000) highlights the importance of reflection to consolidate thinking.

Each task is in the curriculum and also grounded in research. Figure 1.2 (page 14) highlights the importance of including each component in the SNAP to not only address the subskills in number sense but also the five mathematical strands we discuss later in this chapter (see page 19).

Mental models (such as Venn diagrams and flow charts) are effective ways for students to organize their thinking and be engaged in the mathematics classroom (Fuson & Murata, 2007). Researchers Karen C. Fuson, Mindy Kalchman, and John D. Bransford (2005) write when students make connections between conceptual models, their problem-solving ability improves. Another NCTM (2009) study suggests using models is beneficial because they do the following.

- Show how mathematics is connected (that is, in life, people use many different mathematical strands to solve one problem; mathematics is not isolated).

- Show students how to apply mathematics in new and creative ways by combining mathematical concepts.

- Engage students and encourage mathematical reasoning.

- Create an appropriate situation where students must not only cultivate mathematical ideas but also apply them (which motivates students).

- Provide points of entry for students with different levels of mathematical knowledge.

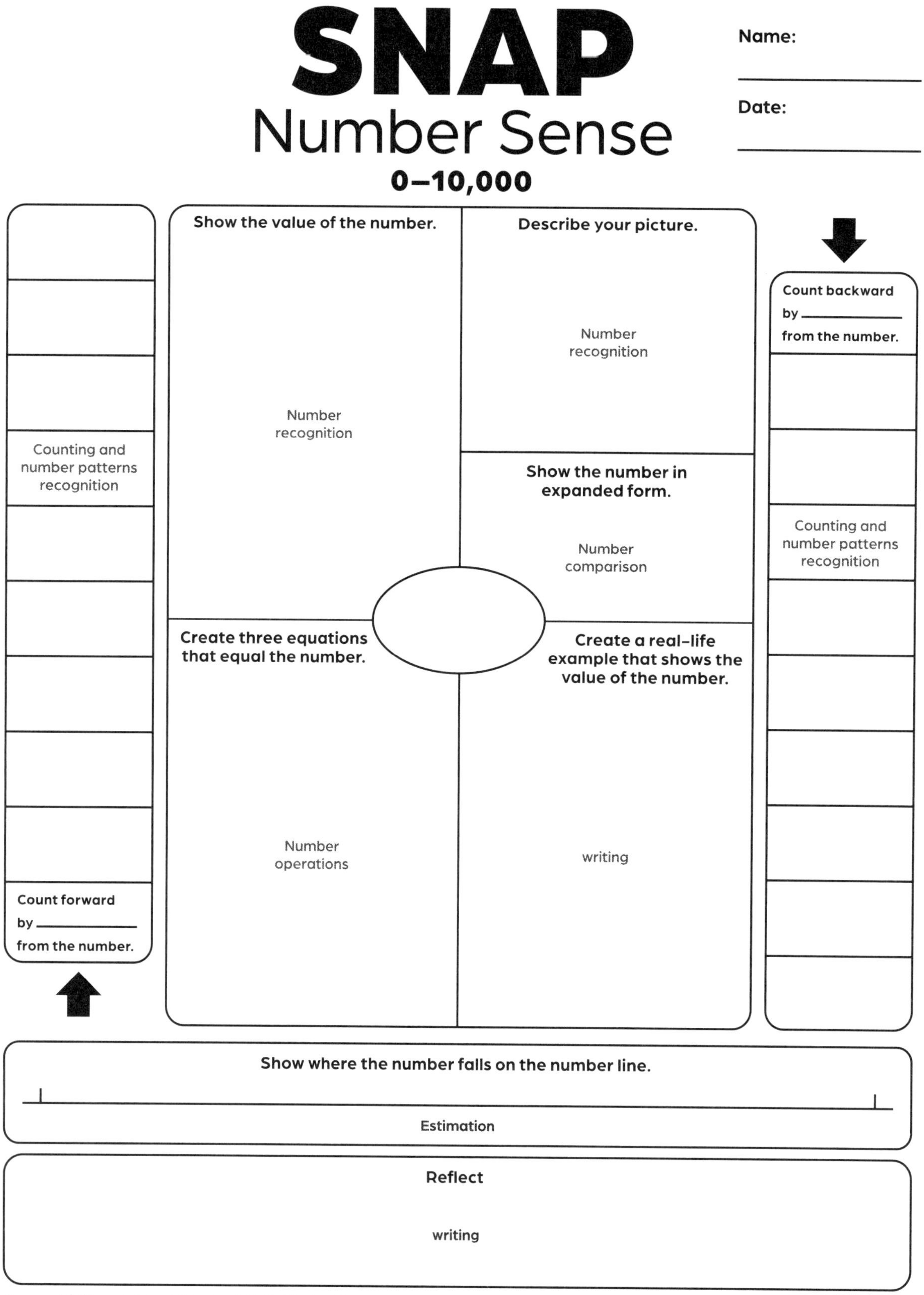

# SNAP
## Number Sense
### 0–10,000

Name: _____

Date: _____

**Show the value of the number.**

Number recognition

**Describe your picture.**

Number recognition

**Show the number in expanded form.**

Number comparison

**Create three equations that equal the number.**

Number operations

**Create a real-life example that shows the value of the number.**

writing

Counting and number patterns recognition

**Count forward by _____ from the number.**

**Count backward by _____ from the number.**

Counting and number patterns recognition

**Show where the number falls on the number line.**

Estimation

**Reflect**

writing

*Source: Chilliwack School District (2016). Used with permission.*

**FIGURE 1.2:** Research underpinnings of the SNAP. The labels correspond to the six primary skills relating to number sense. In addition, the figure highlights areas of writing.

The SNAP is a perfect example of a mental model that not only aids students in organizing and demonstrating their thinking but also incorporates each of these elements in powerful ways.

## THE DESIGN PROCESS

The SNAP works in part because it provides a balance of substance and style by combining competency-based tasks in a game-like format. During the design phase in 2016, the SNAP team delved into various assessments, most notably *The ANIE: A Math Assessment Tool That Reveals Learning and Informs Teaching* by Kevin Bird and Kirk Savage (2014; see figure 1.3, page 16). The team used a backward design approach. Instead of having students solve word problems at the onset, team members created a template for students to do the mathematics first and then create word problems or real-life applications after. The SNAP is similar to the ANIE, but explicitly emphasizes number sense. The ANIE is designed for teachers to assess multiple mathematical concepts (Bird & Savage, 2014). The SNAP team desired to match the appeal of the ANIE's design by creating a one-page, student-friendly document that ignites a sense of wonder while providing maximum accessibility because it requires little reading for students to engage in the tasks.

Researcher Richard Lesh's (1979) translation model also informed the development of the SNAP. Lesh (1979) developed his model-eliciting activity to represent conceptual mathematics, and this understanding aligned with the assessment goals of the SNAP team. Lesh (1979) states:

> In general, "being able to use a concept" involves more than simply "having the concept." Getting an idea into a youngster's head does not guarantee that the new idea will be integrated with other ideas that are already understood, that situations will be recognized in which the idea is relevant, nor that library-type "look up" skills will be available to retrieve related ideas when they are needed. "Being able to use an idea" may also involve specific problem-solving processes in addition to those needed to demonstrate the simple attainment of the concept. (p. 6)

During the design phase, the SNAP team also worked closely with John Mighton, a Canadian mathematician and founder of JUMP Math (https://jumpmath.org/us), as it worked through the complexities of designing a comprehensive assessment. Discussing various approaches to number sense and number sense assessment proved useful as the team continued to refine its thinking. After piloting several draft templates, including one provided by Yarrow Elementary teacher Dawn Motz, the team eventually launched the template in its present format.

# The ANIE

Name: _____ Grade: _____ Date: _____

**Problem:**

**Make your best guess.** _____

**Explain your strategy.** _____

Show how to solve.

| Calculate | Draw a Sketch |
|---|---|
|  |  |

**Explain your calculation and sketch.**

**Give an example of how to use the mathematics in real life.**

**Reflect on your thinking.**   What was easy? What was hard? What did you learn?

*Source: Adapted from Bird & Savage, 2014. Used with permission.*
**FIGURE 1.3:** The ANIE.

## A CLEAR PROGRESSION

With respect to implementation, the creators intended for teachers to use the SNAP to assess specific content areas at each grade level. From a developmental perspective, the SNAP is designed to assess content that gets progressively more challenging. The familiarity of the format allows students to become confident as they encounter numbers that move from two digits to integers. The progression has assigned grade levels (based on standards or typical student learning rates), but the sequence is still effective when students are above or below grade level (see figure 1.4). Teachers use information they glean from the SNAP to inform their instruction, and schools collect the classroom data to guide schoolwide interventions as well. For collecting district data, the formal collection is limited to the minimum outcomes in each grade to ensure the data collection does not overwhelm classroom teachers (see figure 1.5).

| Grade | 2 | 3 | 4 | 5 | 6 | 7 |
|---|---|---|---|---|---|---|
| Number Sense | Number concepts to 100<br><br>Any two-digit number | Number concepts to 1,000<br><br>Any three-digit number | Number concepts to 10,000<br><br>Any four-digit number | Number concepts to 1,000,000<br><br>Any six-digit number | Number concepts from thousandths to billionths<br><br>Any decimal to the hundredths | Integer concepts<br><br>Any two-digit negative whole number |

*Source: Chilliwack School District (2016). Used with permission.*
**FIGURE 1.4:** Content by grade level.

| Content | Item Descriptor |
|---|---|
| Number and computational fluency (number sense) | Demonstrate understanding and application of a decimal fraction through the lens of the curricular competencies (reasoning and analyzing; understanding and solving; communicating and representing; connecting and reflecting). |
| Draw | The picture must show the value of the number. |
| Skip counting | Begin at the number and count forward and backward from the number the teacher chooses. |
| Equations | Students demonstrating full proficiency will use grade-appropriate operations in their equations. |
| Real-life example | The examples must be realistic and specific. It is important that students demonstrate their understanding of value in their example. |
| Number line | Students in grade 6 choose their own end points according to the number the teacher chose for the assessment. Students add benchmarks to the number line to help situate the number. |
| Reflection | Reflections help increase the value of a learning experience and allow students to link ideas and construct meaning from their experiences. |

*Source: Chilliwack School District (2016). Used with permission.*
**FIGURE 1.5:** Content and item descriptors.

## A BASIS IN RESEARCH ON ASSESSMENT

The SNAP is grounded in a variety of assessment models. Educators Richard J. Stiggins, Judith A. Arter, Jan Chappuis, and Stephen Chappuis (2004) developed a set of guidelines for sound classroom assessment (see figure 1.6). Stiggins and colleagues' (2004) research model and the SNAP are congruent in these areas: the SNAP has a clear purpose and clear targets for assessing number sense. Educators can interpret the results with ease, with a focus on student growth over time. Teachers, students, and parents can effectively understand results, and when undertaking a SNAP, student involvement and engagement are high.

| Sound Lesson Design | Descriptors | The SNAP |
| --- | --- | --- |
| Clear purpose | Assessment processes and results serve clear and appropriate purposes. | Teachers use the SNAP *for*, *as*, and *of* learning. |
| Clear targets | Assessments reflect clear and valued student learning targets. | The SNAP links to curricular outcomes and has an accompanying rubric. |
| Sound design | Teachers translate learning targets into assessments that yield accurate results. | The SNAP is a valid and reliable assessment that also involves minimal reading. |
| Effective communication | Teachers manage and effectively communicate assessment results. | The SNAP provides students, teachers, and parents with immediate feedback. |
| Student involvement | Students are involved in their own assessment. | The SNAP allows students to work at their own pace and access any tasks in their own way and in any order. |

*Source: Adapted from Stiggins, et al., 2004.*
**FIGURE 1.6:** Sound lesson design and the SNAP.

Researcher and professor of mathematics education Marian Small (2012) in *Making Math Meaningful to Canadian Students, K–8*, discusses how mathematics assessments need to contain a variety of elements to be effective. Small (2012) explains that a successful assessment plan needs to contain the elements in figure 1.7.

| Effective Elements of Mathematics Assessments | The SNAP |
| --- | --- |
| Balances the measurement of both mathematics content and processes | The SNAP equally assesses mathematics content and processes. |
| Appropriate for its purpose | The SNAP is a visually appealing assessment teachers can use *for*, *as*, and *of* learning. |
| Includes a variety of formats | Teachers can break down the SNAP template into separate elements; it also includes a variety of other templates. |
| Aligns with student needs and expectations | The SNAP is accessible for all students and time is unlimited to complete it. |
| Fair to all students | Teachers can adapt the SNAP to meet the needs of all students. |
| Assists students to assess their own learning | The SNAP provides immediate feedback to students and teachers can use it *for*, *as*, and *of* learning. |
| Measures growth over time | Teachers can use the SNAP throughout the year to measure growth. |
| Sets high, realistic expectations for students | The SNAP is a low-floor, high-ceiling assessment grounded in the curriculum. |

*Source: Adapted from Small, 2012.*
**FIGURE 1.7:** Comparing the SNAP to the elements of successful assessments.

The SNAP contains each of these eight descriptors in figure 1.7. We assert that in both models, the SNAP embodies all the elements required for sound assessment.

## Five Strands of Mathematical Proficiency

Knowing that practical and authentic 21st century assessment extends beyond gathering one-time data, the SNAP design team intends for the assessment to inform the rest of the learning process and drive purposeful, responsive action to guide students toward mathematical proficiency. *Mathematical proficiency*, as defined in *Adding It Up: Helping Children Learn Mathematics* (National Research Council, 2001), "captures what we think it means for anyone to learn mathematics successfully" (p. 5). *Adding It Up* describes five strands of mathematical proficiency to inform mathematics instruction and assessment: (1) conceptual understanding, (2) procedural fluency, (3) strategic competence, (4) adaptive reasoning, and (5) productive disposition (National Research Council, 2001). The SNAP addresses each of these competencies in an interdependent and interwoven fashion. We discuss this more in chapter 3 (page 41).

Remember, *SNAP* stands for student numeracy assessment and *practice*. The term *practice* is essential to the assessment because it follows a spiraled, interweaving approach to teaching and evaluating numeracy skills, according to University of South Florida researchers Doug Rohrer, Robert E. Dedrick, and Sandra Stershic (2015). When teachers use an interweaving approach, they both revisit and deepen their thinking on any given topic, and the learning is spread out over time, not compartmentalized into separate topics. With the SNAP, teachers continuously reinforce the five competencies throughout the year, thus providing students with opportunities to strengthen their understanding. The SNAP is an assessment teachers can repeatedly use (*as* learning) to build student fluency, confidence, comprehension, and skills. Teachers can also use the SNAP as a formative assessment (*for* learning) to inform instructional practices. Finally, teachers use the SNAP as a summative assessment (*of* learning) to report on student achievement. In this sense, the SNAP team developed something quite rare in the world of assessment!

## Unique Features of the SNAP

The SNAP contains many special features that promote student engagement and learning. Traditional mathematics assessments include row after row of questions for students to answer, with students focusing only on their received score. Mathematics assessments are prone to having complex directions or multistep word problems, promoting student confusion and unnecessary complexity. Educators design these traditional assessments to have a right answer with little relevance to students' interests. Finally, traditional mathematics assessments do not include *metacognition* (that is, thinking about your thinking) and are not always culturally sensitive. The following is a list of unique characteristics of the SNAP.

### ASSESSING MATHEMATICS, NOT READING

There are significant connections between reading fluency skills, reading comprehension ability, and performance on word problems in mathematics (Björn, Aunola, & Nurmi,

2016). Because the SNAP focuses on bringing each student's background knowledge to the assessment, there is no requirement for students to read contextual material. Students simply interact with mathematics and do not have to use reading comprehension strategies to figure out the questions. In a 2021 study correlating reading ability to a literacy-heavy numeracy assessment and the SNAP, reading achievement had a significant and moderate effect on the literacy-heavy assessment, and no significant effect on the SNAP. This difference can potentially better assess students who have high mathematics ability and low literacy skills (Savage, 2021). International students, students who are not yet fluent in English, or any students who struggle with reading could potentially demonstrate improved mathematics achievement when using the SNAP because language is less likely to get in the way.

## USING IMAGING STRATEGIES

Taking the abstract and creating concrete models is an essential skill for mathematicians. An effective approach for teachers is to provide students with concrete manipulatives and then connect these concrete examples to symbols and abstract concepts, also referred to as a *concrete, pictorial, and abstract approach* (Bruner, 1966; Putri, et al., 2020; Uttal, Scudder & DeLoache, 1997). The SNAP aligns by starting with an abstract idea (a number), then asking for the student to represent a number (via a drawing), and provide a concrete (real life) example. By creating concrete representations, students can demonstrate their comprehension by applying a concept to a real-life example. These models can be in the form of a drawing, a manipulative, a graph, or even a mental image. Indeed, conceptual understanding is impossible without the ability to image, as imaging is a crucial stage of comprehension (Bird & Savage, 2015; Bower & Morrow, 1990). Additionally, it is helpful when students can demonstrate their work and understanding in an open-ended, multirepresentational fashion. This approach caters to students with different skills, backgrounds, and dispositions, "providing that the result is reasonable, the pathway and mode of representation are valued," as learning experiences (Zevenbergen, Niesche, Grootenboer, & Boaler, 2008, p. 642). Therefore, the SNAP requires students to demonstrate their understanding of the mathematics concepts through drawings, sketches, and representations on the page, which is a *multirepresentational approach.*

## USING OPEN VERSUS CLOSED ASSESSMENT

Open-ended questions in mathematics can support higher achievement levels because educators require students to consider multiple approaches and options to find solutions (Lee, Kinzie, & Whittaker, 2012). A *closed assessment* has a predetermined correct answer. An example of a closed assessment is a multiple-choice test or a mathematics quiz where students calculate the correct answers. When teachers use a traditional closed-mathematics approach to assessment and instruction, students primarily develop procedural knowledge, but this procedural knowledge is of limited use to students in novel situations (Boaler, 1998). Coauthors Jo Boaler, Jen Munson, and Cathy Williams (2021) lean further into the importance of open-ended questions and coined the learning phrase "low floor, high ceiling." They define *low floor, high ceiling tasks* as tasks where "everyone can engage, no matter what his or her prior understanding or knowledge, but one that is open enough to extend to high levels, so that all students can be deeply challenged" (Boaler et al., 2021, p. 2).

The SNAP is an opened-ended assessment. *Open-ended questions* "induce more cognitive strategy" use than closed questions, such as multiple choice or right-or-wrong answers (O'Neil & Brown, 1998). A three-year study involving students from two schools finds students who are taught regularly from a pedagogy of open-ended activities perform better than students taught in a more traditional, textbook-based approach (Boaler, 1998). A similar study by coauthors Hamide Çakır and Özge Cengiz (2016) also shows improved student engagement and achievement when educators used open-ended questioning techniques in elementary classrooms.

The SNAP begins with a conceptual question called a *doorway question* (Bird & Savage, 2014). The doorway question (which is the number in the middle of the SNAP template) relates to the concept the teacher is assessing. For instance, if teachers are assessing number concepts to 100 on the SNAP, the doorway question should be a developmentally appropriate number for each student between 0–100. Students use their own background knowledge and experiences to explain their understanding to answer the question.

## MAKING REAL-LIFE CONNECTIONS

Helping students make real-life connections to mathematics is another foundational piece of mathematical proficiency, enhancing student understanding of concepts, motivating learning, and helping students apply mathematics to authentic problems (Gainsburg, 2008). Relevancy matters in helping students make connections. Mathematics has long been the only content area where accessing large amounts of fruits and vegetables seems to be the norm. You're probably familiar with the old mathematics textbook questions that read something like the following.

> **Tommy brought 72 kumquats to school to share with his 12 friends. How many kumquats did each friend receive?**

The lack of connection for students in this question causes them to "chase all the mental squirrels" popping up in their heads. "What's a *kumquat*? It's a funny word! I like saying it. How did he get the kumquats to school? Why does he think his friends want kumquats? If this is Tommy in my class, he doesn't have twelve friends anymore because Billy said he doesn't like Tommy." The mathematics is lost!

In a literacy context, consultants and coauthors Ellin Oliver Keene and Susan Zimmerman (2007) conclude the best approach to real-life connections is to help students make text-to-text, text-to-self, and text-to-world connections. These connections are crucial for student engagement and understanding and also a way to address confidence and self-efficacy (Stoehr, Turner, & Sugimoto, 2015). In a similar manner, mathematics students should also make real-life connections: "The pedagogical practice of connecting mathematical content to real-world contexts, particularly contexts relevant to students' knowledge and experiences, can positively impact student motivation as well as promote conceptual understanding" (Sugimoto, Turner & Stoehr, 2017, p. 61). The SNAP requires students apply the mathematics concept from the assessment to a real-life context. This process helps move the mathematics concept from abstract thinking to concrete action.

**IMPROVING METACOGNITIVE STRATEGIES**

Teaching students how to improve their metacognitive skills significantly impacts their understanding and achievement in all teaching domains, including mathematics (Hattie, 2009). The term *metacognition* refers to an awareness and ability to control your thinking processes; another definition is *thinking about your thinking*. In multiple studies, researchers Gökhan Öszoy (2011), and Öszoy and Ayegül Ataman (2009) find grade 5 students explicitly taught metacognitive strategies performed significantly better in mathematical problem-solving achievement than control groups that received no training. In another study, Kelli R. Thomas (2006), an assistant professor in the department of teaching and leadership at the University of Kansas, finds students who were explicitly taught metacognitive skills show enhanced abilities to complete problem-solving processes. Thomas (2006) states, "The ability to think about, and reflect on, the problem-solving process is an important component of learning mathematics" (p. 86). The SNAP requires students to reflect on their thinking and learning at the end of the assessment. Prompts for students include "What was easy?" "What was difficult?" and "What do I still need to learn?" (Chilliwack School District, 2016).

The alternative to metacognition is to *not* think about your thinking and simply get to the answer. Mathematics is a content area where it is tempting to say there is only one correct answer and define *mathematics* as a very dichotomous (that is, correct or incorrect) subject. But this idea strips away the wonder of mathematics, which resides in getting to the answer. While it is strictly true that there is sometimes only one correct answer, the pathways to that answer are what make mathematics a subject of discovery. Here's a very simple example: When asked, "What is ⅓ of 12?" The student immediately responds, "4." In the right or wrong dichotomy, you credit them with being correct. The wonder comes in asking how they got the answer, to which they replied (with a great deal of satisfaction), "Those are so easy! I added the 1 and the 3 in the fraction." They got the correct answer, but their strategy is flawed. If you ignore the process and only focus on the product, you miss a wonderful and wonder-filled teaching and learning opportunity. The response also cannot be to simply drill the rule into the student (in this example, *of* means to multiply, and you need to make the whole number a fraction). Instead, and to encourage the exploration, you explain the student's method is not foolproof. For example, ⅔ of 12 is not 5, but 8. Encourage the student to be metacognitive *and* reflect on their work *and* potentially interact with peers, who may hold another piece of the puzzle.

**EMPLOYING CULTURAL SENSITIVITY AND COMPETENCE**

Oregon State University emerita faculty member Jean Moule (2012) defines *cultural competence* in an educational setting as "the ability to successfully teach students who come from cultures other than your own" (p. 5). Maya Angelou, poet and civil rights activist, is credited with the quote, "I did then what I knew how to do. Now that I know better, I do better" (Goodreads, n.d.). Angelou's sentiment is playing out in schools across British Columbia, where educators focus on improving equity and inclusion for all students. As author, thought leader, and strategist Glenn E. Singleton (2015) says, "As schools work toward equity, they will narrow the gaps between the highest- and lowest-performing groups and eliminate the racial predictability regarding which groups achieve in the highest- and lowest-performing categories" (p. 6).

By design, the SNAP is *culturally sensitive* because it requires students create their own real-life connections with the mathematics concepts. These authentic connections mean the assessment does not impose an unfamiliar mathematical situation or story that requires student background knowledge to access the context. Students can think in their native language to present their understanding of mathematical concepts and learning—an important principle of creating equitable practice in diverse classrooms (Zevenbergen et al., 2008).

### REDUCING MATHEMATICS ANXIETY

Maureen Finlayson (2014), an author and assistant professor at Cape Breton University in Sydney, Nova Scotia, Canada, asserts several factors cause student mathematics anxiety, including strict adherence to fixed curriculum, a reliance on closed (right or wrong) assessment practices, and a teaching approach that promotes outcomes over process. Finlayson's (2014) research supports a teaching approach that values process over product and relies on a deeper dive into conceptual mathematics understandings to overcome student mathematics anxiety in classrooms. *Number talks* (also referred to as *mathematics talks*) are strategies teachers use to improve student learning in mathematics classrooms by focusing on mathematical processes and reducing mathematics anxiety (Boaler, 2014). Instead of worrying about answering fifty questions on a timed test, students spend their learning and thinking time on one question at a time during a number talk. Much like number talks, the SNAP supports a deep dive into one question. The SNAP is just one page and uses a graphic organizer and scaffolded approach. One question may also not be as daunting for proficient students by producing less test anxiety and ensuring they have time to show more accurately what they know (Bird & Savage, 2014).

## Classroom Applications

As we outlined earlier (see page 5), the SNAP is an assessment teachers can use *for*, *as*, and *of* learning. For teachers, it is a very useful tool to find a student's strengths and areas of growth. When interpreting a completed SNAP, themes will emerge, allowing teachers to address issues for a whole class, a small group, or an individual. Students quickly see their own strengths and areas to work on *for* learning. When students are interacting with the SNAP, they are making a plethora of connections and deepening their own learning, so the SNAP is a valuable tool for students to use *as* learning. Finally, teachers can use the SNAP summatively (or *of* learning) to measure benchmarks throughout the year.

The student example in figure 1.8 (page 24) reveals a great deal of information about each set of tasks. The teacher asked students to count forward by 10, which is relatively

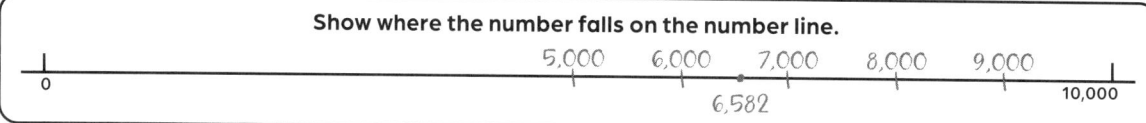

**FIGURE 1.8:** Number sense student example (grade 4).

easy, while counting backward by 5 is more challenging. The symbols for the picture and the justification show the student clearly understands this area, and the representation of the expanded form is accurate as well. The three equations are simplistic; this is an area where the teacher could encourage the student to apply more complexity in the future. The real-life example is accurate and contextually correct; however, this is also an area where teachers can stretch the student's thinking. The number line task, which includes midpoints, reveals the student understands this area. The student displays a great deal of thinking in this example, providing rich information for both the learner and teacher alike.

Many teachers experience challenges concerning teaching mathematics, particularly elementary teachers. Elementary teachers are not subject specialists like their high school counterparts; school leaders require them to teach many subjects. Mathematics anxiety is also very commonplace, and elementary teachers are not immune to this fact:

> The body of research that has resulted in findings that elementary school teachers experience math anxiety, convey negative perceptions of mathematics, possess low levels of math knowledge and teaching methods, and express low levels of confidence in their abilities to teach mathematics provides reasons that may influence teachers' attitudes to avoid or lack interest in teaching mathematics. The impact of these attitudes can be influential factors in low student scores on standardized mathematics achievement tests. (Porter, 2019, p. 10)

How many elementary teachers choose mathematics as their most confident subject? In our experience, very few. It makes sense for many teachers to ask, "Where do I begin? How do I assess my students in mathematics? How do I get my students to think deeply about numbers? How do I teach for understanding? How do I differentiate for the diverse needs in my classroom?" The SNAP assists teachers in answering these questions with respect to number sense by providing an engaging road map they can use to measure their students' abilities. The SNAP requires very little advanced understanding and preparation; it is easily accessible for students and provides rich and timely information.

The SNAP also benefits teachers by offering them multiple implementation options. Using the SNAP encourages students and teachers to think about mathematics in novel and unique ways, promoting the concept of assessment as a means of learning and growth, thus providing a practical approach for educators. Instead of teachers asking students to perform isolated tasks, they integrate the challenges on one page, enabling students to see many connections and interrelationships when working with numbers.

## USING THE SNAP AS A FORMATIVE ASSESSMENT

Teachers can use the SNAP any time of the year in various ways, depending on the intent. During school start-up, teachers primarily use the SNAP as a formative assessment to discover students' strengths and weaknesses. As students move through different grade levels, they will likely become well acquainted with the SNAP. It is important for teachers to approach the SNAP with creativity and enthusiasm, rather than treating it like just another worksheet. To keep things fresh and engaging, try incorporating unique techniques such as *zooming in* (picking one area of the SNAP template to focus on) or experimenting with some of the different themed templates we share in chapter 5 (page 71).

### USING THE SNAP AS A SUMMATIVE ASSESSMENT

Teachers can use the SNAP as a whole-class activity in multiple ways. They can assign all students in a class the same number for a summative assessment. Using the SNAP in this way, the assessment lends itself to formal reporting, such as report cards or data collection.

Each of the five strands of the SNAP (conceptual understanding, procedural fluency, strategic competence, adaptive reasoning, and productive disposition) has a maximum of 4 points, totaling a score of 20. Here are the recommended cut scores.

- 0–7: Emerging
- 8–12: Developing
- 13–16: Applying
- 17–20: Extending

Using these cut scores, teachers can quickly assign an overall summative grade and performance for each completed SNAP. These scores translate easily for school or district data collection and analysis.

Alternatively, teachers can assign different numbers to individuals or small groups, enabling both differentiation and equity. Teachers may want to focus independently on a task like the number line or expanded form as opposed to the whole document.

### USING THE SNAP FOR SMALL GROUPS

The SNAP is particularly beneficial in small-group settings, where teachers can work closely with teams of students to guide discussions about their approach to the assessment. Through this collaborative process of peer-to-peer conversation, students can learn new and innovative ways to address each challenge under the careful guidance of their teacher. This approach fosters the development of fresh ways of thinking, and students can quickly identify and address errors in a supportive manner. When teachers use the SNAP formatively, they glean information about each student, what the student can and cannot do, and the student's current limitations. Invariably, only some students can show their understanding using pencil and paper. When spending time one to one with each student after an assessment, teachers can better understand each student's thought processes, and why each student may have struggled to demonstrate understanding. Teachers can analyze each student's learning by identifying the areas in which the student excels, and those areas the student finds challenging. Essentially, teachers using the SNAP can determine whether each student is following a formula or whether each student is truly grasping the underlying concepts.

### HELPING STUDENTS WHO FAKE SKILLS

*Faking it* is a real problem in schools, and it happens more often than you might expect. Whenever teachers implement the SNAP in a classroom, one of the most important pieces of information it provides is the identities of those students who have poor numeracy skills and highly developed coping skills. These students are *survivors* who have figured out how to muddle their way through the crowded classroom curriculum. Common strategies for survivors include the following.

- Working in small groups or with a partner
- Copying the work of other students
- Completing any assigned work at home (with the help of a parent)
- Asking for bathroom breaks
- Explosive behaviors
- Nonattendance

When teachers first introduce the SNAP to a classroom full of students where competency-based assessments are uncommon, the survivor students come to the surface and can receive the help they need. Because the SNAP is a personalized, open assessment, students are unable to copy their responses. They must interact with the mathematics and respond. Teachers become quickly aware of whether students are on track or need help. The best analogy takes place in a reading classroom when a teacher listens to a student read, one to one. There is no hiding for the student. The student can either do it or can't, and if the student can't, it is very clear where support is required.

### USING THE SNAP FOR STUDENTS WITH INDIVIDUAL EDUCATION PLANS

The SNAP can also be valuable for students on individualized education plans because the reading requirement, which can be a stumbling block for many learners, is limited. Also, the SNAP allows teachers to choose any number or equation that best matches the ability level of a student. Additionally, learning assistance teachers and resource teachers can use the SNAP with small groups of students without embarrassing them. Knowing *all* students use the SNAP allows for a sense of normalcy, inclusivity, and equity.

## Chapter Summary

This chapter discusses why the SNAP works and how it relates to the five strands of mathematical proficiency. It discusses the unique features of the SNAP, such as reducing mathematics anxiety, helping students make real-life connections, and more. Finally, this chapter gives examples of using the SNAP in a classroom. In chapter 2 (page 29), we provide background on the concept of number sense and its importance in the classroom. We demonstrate how kindergarten, elementary (grades 1–5), and middle school (grades 6–8) teachers perceive and view number sense. Further into the book, we describe how each skill within the SNAP relates to the five strands of mathematics proficiencies (National Council of Research, 2001). In chapter 2 (page 29) we continue to reveal how teachers can use the SNAP as a foundational assessment tool in their classrooms.

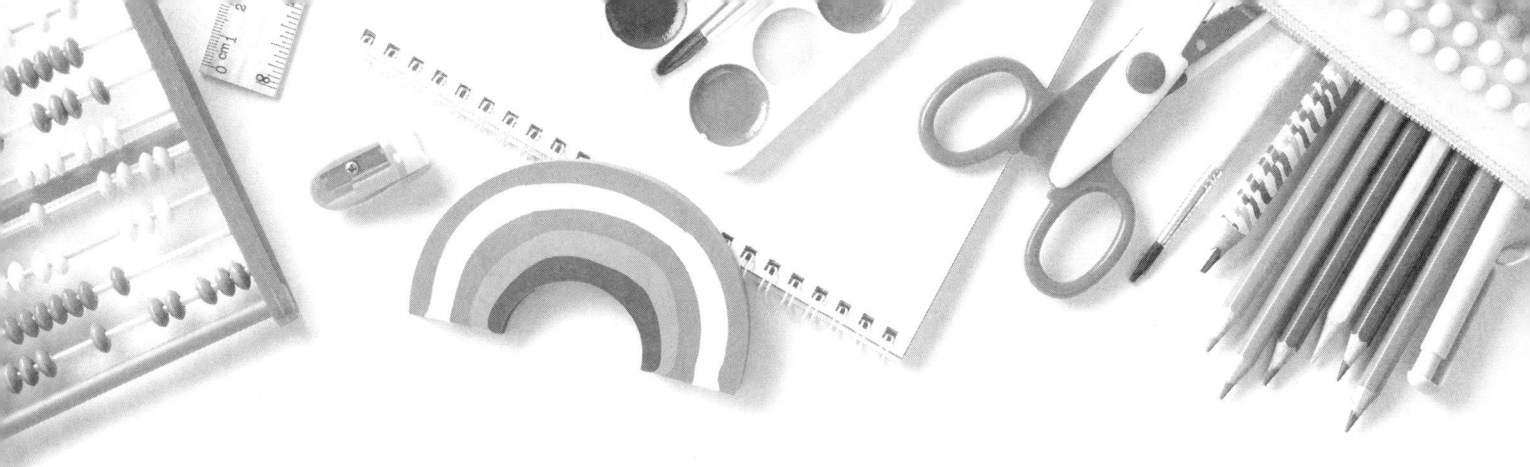

# CHAPTER 2

# Looking at Number Sense Foundations

Number sense begins when students are very young. They ask, "How many are there? How much more? How much longer until we're there?" Parents encourage their children to count on their fingers and toes, and sing songs with repeated patterns. Sometimes, parents notice children playing with their food by putting it into small groups. Also, toddlers quickly pick up concepts such as asking for more food or being told "that's enough" during mealtime. They hear their parents say, "One more spoon of vegetables and then dessert," or "Finish what's on your plate before playtime." When students are at school, their teachers commonly say, "One, two, three look at me," or "Three more minutes and then we're done." Or teachers count down from five to bring everyone to attention in their classrooms.

Students enter school full of curiosity and wonder, especially concerning numbers. They love to count things off and sort them, and they are fascinated with patterns. They use manipulatives to explore and demonstrate their thinking. When they are very young, students begin to put together the building blocks of number sense, laying a foundation for mathematical understanding. We cannot stress enough the importance of number sense instruction in kindergarten. Researchers and coeditors Christopher T. Cross, Taniesha A. Woods, and Heidi Schweingruber (2009) support the belief that number sense is of utmost importance in kindergarten. Researcher Zuhal Yilmaz (2017) explains, "the reasons why children face difficulties in learning mathematics in later grades . . . might be due to students' underdeveloped number sense in early elementary school grades and kindergarten" and "that developing early number sense plays a critical role in laying the foundation for children's future academic achievements in mathematics" (p. 892). Citing the works of Amal Alajmi and Robert Reys (2007) and Nancy I. Dyson, Nancy E. Jordan, & Joseph Glutting (2013), Yilmaz (2017) posits that in addition to this critical role, having enough opportunities and support to learn number sense skills in the early years allows students to build a strong background for learning advanced mathematics.

Because number sense is a core building block in the acquisition of mathematical thinking for students, there is an opportunity to improve how teachers assess and teach this core competency using the SNAP. The remainder of this chapter highlights the importance of number sense from a theoretical perspective, including examining grades K–8, where we talk about what teachers can accomplish when using the SNAP as a teaching tool. We also highlight the ongoing use of the SNAP through the lens of teacher practitioners—those heroes working daily with students.

## A Theoretical Perspective on Number Sense

Judith Sower, former chair of the NCTM Research Advisory Committee, defines *number sense*:

> As students work with numbers, they gradually develop flexibility in thinking about numbers, which is a hallmark of number sense. . . . Number sense develops as students understand the size of numbers, develop multiple ways of thinking about and representing numbers, use numbers as referents, and develop accurate perceptions about the effects of operations on numbers. (as cited in NCTM, 2000, p. 80)

In her 2007 book *About Teaching Mathematics*, educator Marilyn Burns describes number sense this way:

> Students with good number sense can think and reason flexibly with numbers, use numbers to solve problems, spot unreasonable answers, understand how numbers can be taken apart and put together in different ways, see connections among the operations, figure mentally, and make reasonable estimates. These students seem to have good intuition about numbers and see numbers as useful. (p. 24)

Educators and coauthors Alistair McIntosh, Barbara J. Reys, and Robert E. Reys (1992) provide this definition of number sense:

> Number sense refers to a person's general understanding of numbers and operations along with the ability and inclination to use this understanding in flexible ways to make mathematical judgements and to develop useful strategies for handling numbers and operations. It reflects an inclination and an ability to use numbers and quantitative methods as a means of communicating, processing and interpreting information. It

> results in an expectation that numbers are useful and that mathematics has a certain regularity. (p. 3)

Historically, there have been many challenges concerning numeracy assessment. The National Research Council (2001) states:

> To ensure that students are meeting standards, states and districts have, during the past decade or so, mandated a variety of assessments in mathematics, many with serious consequences for students, teachers, and schools. Although intended to ensure that all students have an opportunity to learn mathematics, some of these assessments are not well aligned with the curriculum. Those that were originally designed to rank order students, schools, and districts seldom provide information that can be used to improve instruction. (p. 4)

Research posits that number sense is a foundational skill in mathematics, inherent from birth and learned early:

> Approximate number sense is believed to be an inborn set of neurological abilities common to humans and some animals. Early number sense includes learned skills involving explicit knowledge, such as counting items using number words and comparing numbers represented symbolically as numerals. (Whitacre et al., 2020, p. 104)

The learning progressions in the *Common Core State Standards for Mathematics* also suggest that an emphasis on number sense development in grades K–2 is a key component of an effective mathematics instruction and assessment model, which serves as a critical foundation for all subsequent mathematical progress (National Governors Association Center for Best Practices [NGA] & Council of Chief State School Officers [CCSSO], 2010b).

## Number Sense in the Early Years: Kindergarten

In kindergarten, it is vital for teachers to establish foundational understandings that connect to real-life applications. These skills set the stage for students' success in mathematics as they progress through each grade. Number sense in the early years is not only memorizing and simple recall but also developing an overall understanding of numbers and their corresponding quantities. While there is a place for rote learning, understanding *how* to decompose a number such as 5 into 3 and 2, 4 and 1, or 5 and 0 (*part-part-whole*) makes all the difference in applying and transferring that knowledge. Learners know they get it when they can subitize all manner of things, such as dice, ten frames, *rekenreks* (a counting rack that contains beads), dots, fingers, and so on. Students scaffold new learning faster when they understand the *why* behind the numbers (see figure 2.1, page 32).

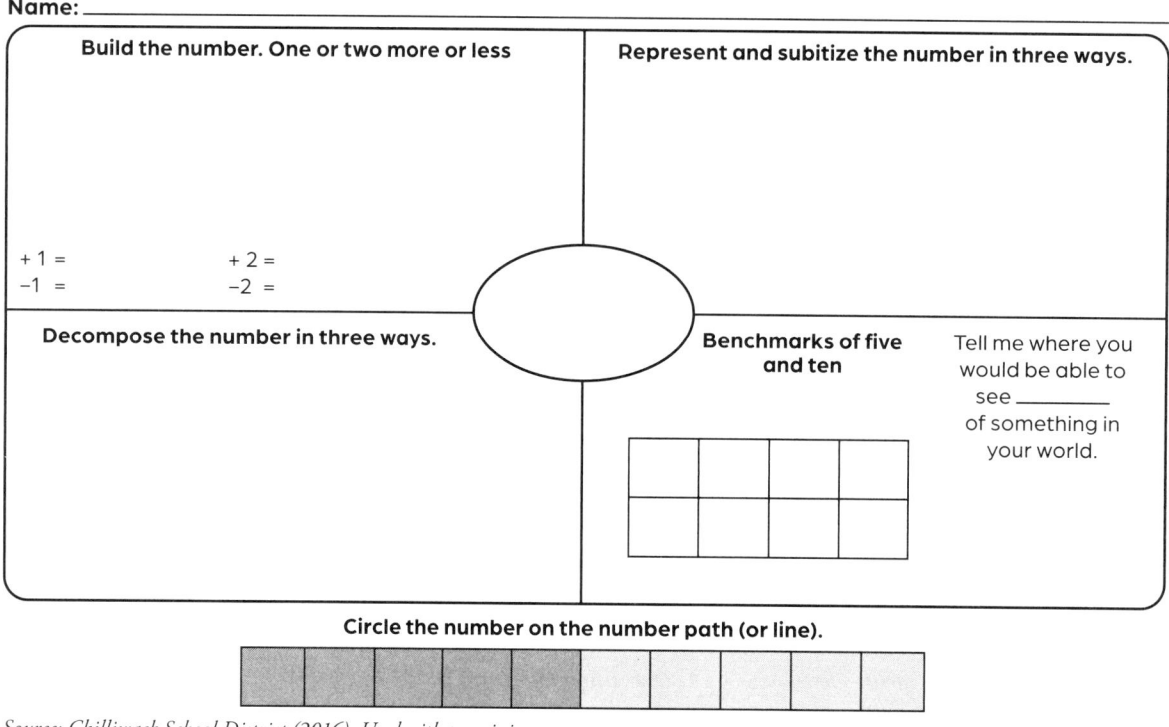

**FIGURE 2.1:** Kindergarten number sense SNAP.

Source: Chilliwack School District (2016). Used with permission.

Visit **go.SolutionTree.com/mathematics** for a free reproducible version of this figure.

The teacher gives all kindergarten students a number and asks them to work through a series of tasks: build the number with one or two more or less, represent and subitize the number in three ways, decompose the number in three ways, demonstrate benchmarks of five and ten, tell where they would be able to see the number in the world, and circle the number on a number path (or line).

Kindergarten teachers believe an effective way of teaching mathematics is not just having a stand-alone mathematics lesson but also incorporating mathematics concepts into instruction and exploration throughout the school day. Nature walks are practical and engaging, and students can collect quantities (leaves, stones, flower petals, or pinecones), which they then connect back to the lesson. Units on measurement can take place entirely outdoors; have students use sticks and cones to introduce length and height. The sandbox is an accessible location to explore capacity, and students can test the concept of weight on swings or with rocks. Additionally, students can explore number sense through art, calendar time, story time, and physical education class. Students understand number sense when they can apply what they know to non-mathematical situations, explain their thinking, and tell how or why they got to the solution.

In kindergarten, beginning with one concept at a time, teachers gently guide students through a series of tasks to demonstrate their understanding of number sense. Using scaffolds such as manipulatives, stories, and technology, students work through different aspects of the SNAP in whole- and small-group settings. We provide a detailed explanation of this process in chapter 4 (page 61).

> **PRACTITIONER'S PERSPECTIVE**
>
> Veteran kindergarten teacher Nancy Ferris (who teaches at Promontory Heights Elementary in Chilliwack, British Columbia, Canada) explains that the challenge of assessment is balancing the curriculum's multiple demands with what she knows to be good pedagogy. Traditional assessment practices place a premium on surface-level understanding and miss the deeper learning connections.
>
> Nancy believes using real-life situations is the most effective way to teach number sense in the classroom; this approach enables students to apply their existing knowledge and derive meaning from a situation. She uses the SNAP to establish a connection with real-life situations from the very beginning of the lesson. She asks students to count everyday items in the classroom, such as crayons, pompoms, or boots.
>
> Ultimately, Nancy believes that kindergarten provides so much learning through play (by allowing curiosity and discovery to lead the students), which results in more deeply engaged learners. The SNAP helps her stay focused on the thinking part of the curriculum and provides a simple yet accurate demonstration of student learning—or *learning artifact*—perfect for use in parent conference updates or report cards.

## Number Sense in Grades 1–5

Not only is teaching number sense essential in kindergarten but also in grades 1–5. Writer and journalist Lauran Neergaard (2013) states, "Those who lagged behind their peers in a test of core math skills needed to function as adults were the same kids who'd had the least number sense or fluency way back when they started first grade." In Chilliwack, several teachers explain that classrooms are diverse and therefore, teachers must be creative and responsive in delivering curricular competencies and pivoting to the most effective direction. They believe students must be confident in verbally communicating their mathematical strategies. Teachers favor number talks, peer-led group work, or targeted mathematics learning centers for students to demonstrate confidence and proficiency. An effective teaching mode is *teacher as facilitator*, where students work in groups, exploring mathematics concepts at their developmental level, and explain their understanding of the mathematics concepts through guiding questions. Engagement is high because all students have an access point to conceptualize the competencies, such as through the computation of numbers or a mathematics story they create using manipulatives. Teachers connecting with their students fosters their natural curiosity, enabling them to aspire to great heights.

As students progress through grades 1–5, they build confidence with number sense and discover more as they interact with the SNAP. Students in these grades are becoming familiar with school routines, so they begin to explore more abstract concepts and gain more real-world mathematics experiences. Student independence is greater in these grades, and students become more adept at collaborating with partners and group members. Manipulatives are common, and teachers expect more concerning fluency and comprehension. In these grades, students begin to see the relationships between addition, subtraction, multiplication, and division (see figure 2.2).

In grade 2, the teacher gives students a number and asks them to work through a series of tasks (draw to represent the number, write the number in expanded form, create three equations that equal the number, write a real-life example that shows the value of the number, and show where the number belongs on the number line). Then, students share their reflections.

Teaching professor Valerie N. Faulkner (2009) explains that algebraic and geometric thinking, form of a number, base ten, equality, numeration, quantity and magnitude, and proportional reasoning are essential to number sense. Through discussions, guided practice, and manipulatives, students in grades 2–5 expand their mathematical fluency and comprehension based on the solid foundation of number sense built in kindergarten and grade 1. The SNAP provides teachers with a rare window into the mathematical thoughts of students to ensure *all* students start with a strong foundation of number sense at an early age.

When students interact with the SNAP in grades 1–5, teachers should expect more than the students demonstrated in kindergarten. Academic rigor begins to look different as students shift away from play-based approaches to more teacher-directed approaches to learning. Teachers expect students to interact with larger numbers and justify their thinking in more complex ways. Despite this shift in expectations, when using the SNAP in grades 1–5, a spirit of inquiry is essential to maintaining high levels of engagement. Create environments that inspire discourse and curiosity, such as having students work in pairs (or small groups), encouraging risk taking, and welcoming mistakes. It is also important for teachers of these grades to avoid the pitfall of misusing the SNAP as just another worksheet. We have seen examples of *SNAP fatigue*, or where students groan or roll their eyes if they view the SNAP as another worksheet. We recommend using the SNAP judiciously, focusing on certain aspects (or *SNAPshots*), and employing other templates (see more about this process in chapter 5, page 71).

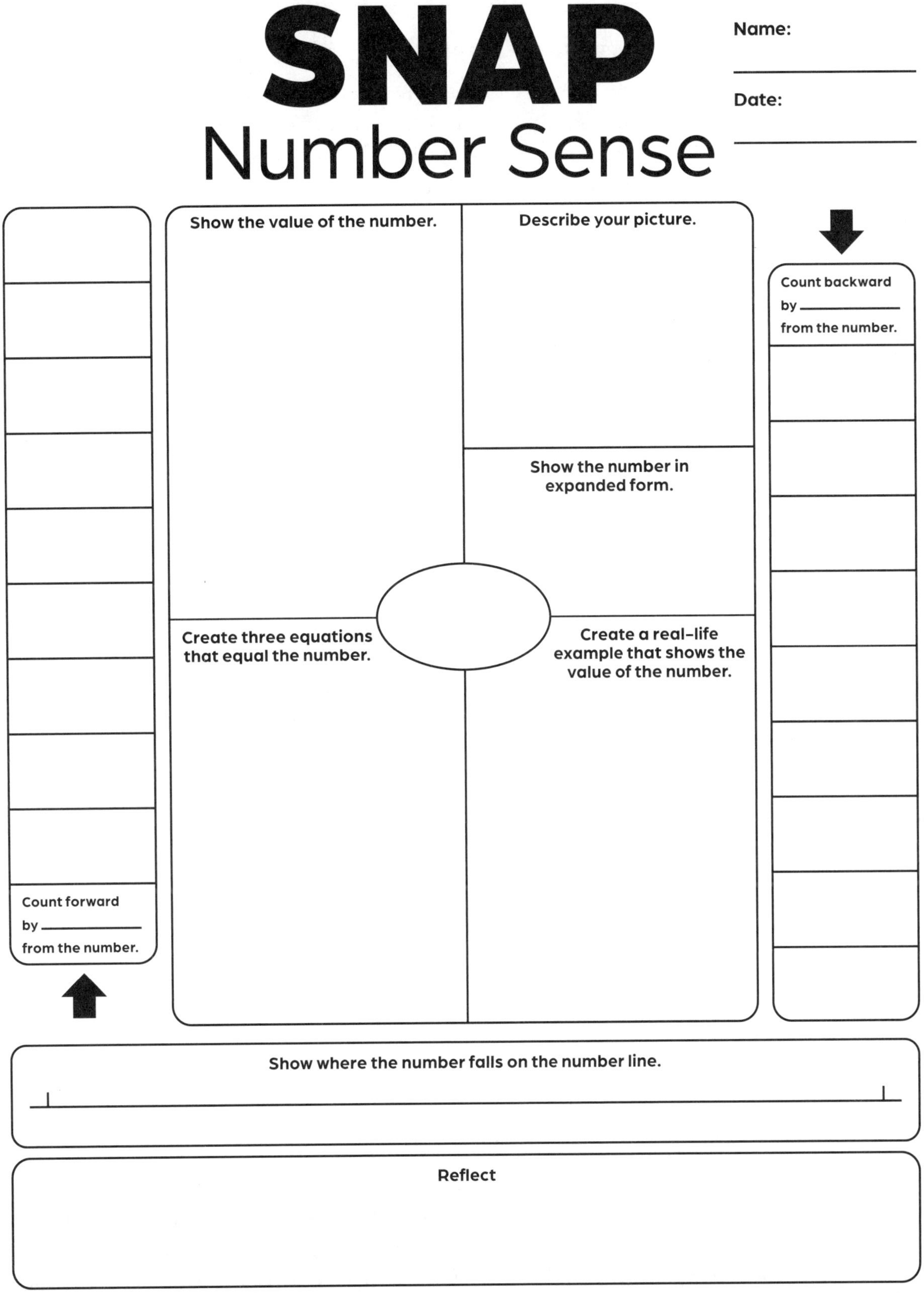

**Looking at Number Sense Foundations**

# SNAP
## Number Sense

Name: _____

Date: _____

**Count backward by _____ from the number.**

**Show the value of the number.**

**Describe your picture.**

**Show the number in expanded form.**

**Create three equations that equal the number.**

**Create a real-life example that shows the value of the number.**

**Count forward by _____ from the number.**

**Show where the number falls on the number line.**

**Reflect**

Source: Chilliwack School District (2016). Used with permission.

**FIGURE 2.2:** Number sense SNAP template.

*Visit **go.SolutionTree.com/mathematics** for a free reproducible version of this figure.*

> **PRACTITIONER'S PERSPECTIVE**
>
> Anna Webb is a Chilliwack School District numeracy teacher who has taught extensively (and successfully) at both the primary and intermediate grade levels. She has spent many years teaching in high-needs, inner-city classrooms, where her students performed in the top 10 percent of all students in the Chilliwack School District on the grade 4 provincial Foundational Skills Assessment. She explains that number sense is essential for students because it is the foundation of all mathematical concepts. As an original member of the SNAP creation team, Anna believes when students have number sense, they gain confidence and will start to experiment with number combinations. Her aim is to have students show flexible thinking and resilience with mathematical computations. She stresses that when early learners have a strong number sense, they can push through the textbook answer and become artists of innovative thinking, which they can demonstrate (for example) in the imaging aspect of the SNAP. Anna explains that students have a firm grasp of number sense when they strongly understand place value. When students can take a number and decompose it with many different operations, it shows they are confident in their knowledge of number value. Teachers foster this confidence through peer-led group work, which focuses on student conversations, small-group support (with teacher guidance), and one-to-one interventions to address foundational learning gaps.

## Number Sense in Grades 6–8

During adolescence, students encounter a plethora of physical, emotional, and mental changes. They also transition to middle school, where new opportunities for learning unfold, along with opportunities to build new networks. The teacher must consider the students' prevalent need to belong and connect with others (including on social media). At this point in their schooling, students will have encountered the SNAP many times, resulting in both opportunities and challenges. On the one hand, they are familiar with the template, and on the other hand, this sense of familiarity can hinder curiosity. Knowing there are other templates (see appendix A, page 117), leaders encourage teachers to embed the strengths of the SNAP in novel ways. We believe in the SNAP's uniqueness in bringing out the best thinking of all students, including those in the grades 6–8.

Based on the learning outcomes in grades 6 and 7 (see figure 1.4, page 17), notice the level of complexity in the SNAP increases yet again and how important it is for students to remain curious about mathematics. Students' understanding of number sense deepens as they progress through these grades, and students become aware of even more complex interrelationships between numbers and make connections with number sense in other mathematical areas.

When teachers focus on number sense in the middle school years, specifically in a way that students pull apart numbers and put them back together in different forms, make reasonable estimates, create and manipulate mental images, and apply it all to the world around them, this is where students find the real joy of mathematics—in unlocking mysteries and solving puzzles.

When students are in their middle school years, educators desire their number sense skills to advance even more, enabling students to handle larger numbers and engage in more abstract thinking. With a strong foundation in base ten and regular practice, students will gain confidence when approaching new challenges. As the SNAP template becomes familiar, the opportunity for teachers to push students' creative and flexible thinking increases. Therefore, lean into this as an educator and, once again, resist the temptation to use the SNAP as just another worksheet. Establish an entry point for each of your students. This is critical; some students in this age group will be at risk of becoming disenfranchised with mathematics learning. Know that the value of promoting inquiry and curiosity can make the difference between engaging or losing a young mind.

Using a sample SNAP at grade 6 (see figure 2.3, page 38), the teacher gives students a number and asks them to work through a series of tasks (draw to represent the number, write to describe the picture, write the number in expanded form, create three equations that equal the number, write a real-life example that shows the value of the number, and show where the number belongs on the number line). At the conclusion, students write a reflection on their learning.

Assessing students throughout the year is important in all grades, and the SNAP's role remains pivotal in grades 6, 7, and 8. During these years, many students develop the capacity to complete the SNAP independently. Once a SNAP is complete, teachers can readily identify areas of strength and challenge and record the information on a class profile (see appendix A, page 117) to explore themes and plan for whole-class and small-group instruction (see appendix B, page 129). Some students may need to be retaught concepts introduced in kindergarten and grade 1, and grades 2–8 teachers are encouraged to collaborate with their primary colleagues regarding number sense pedagogy in the early years.

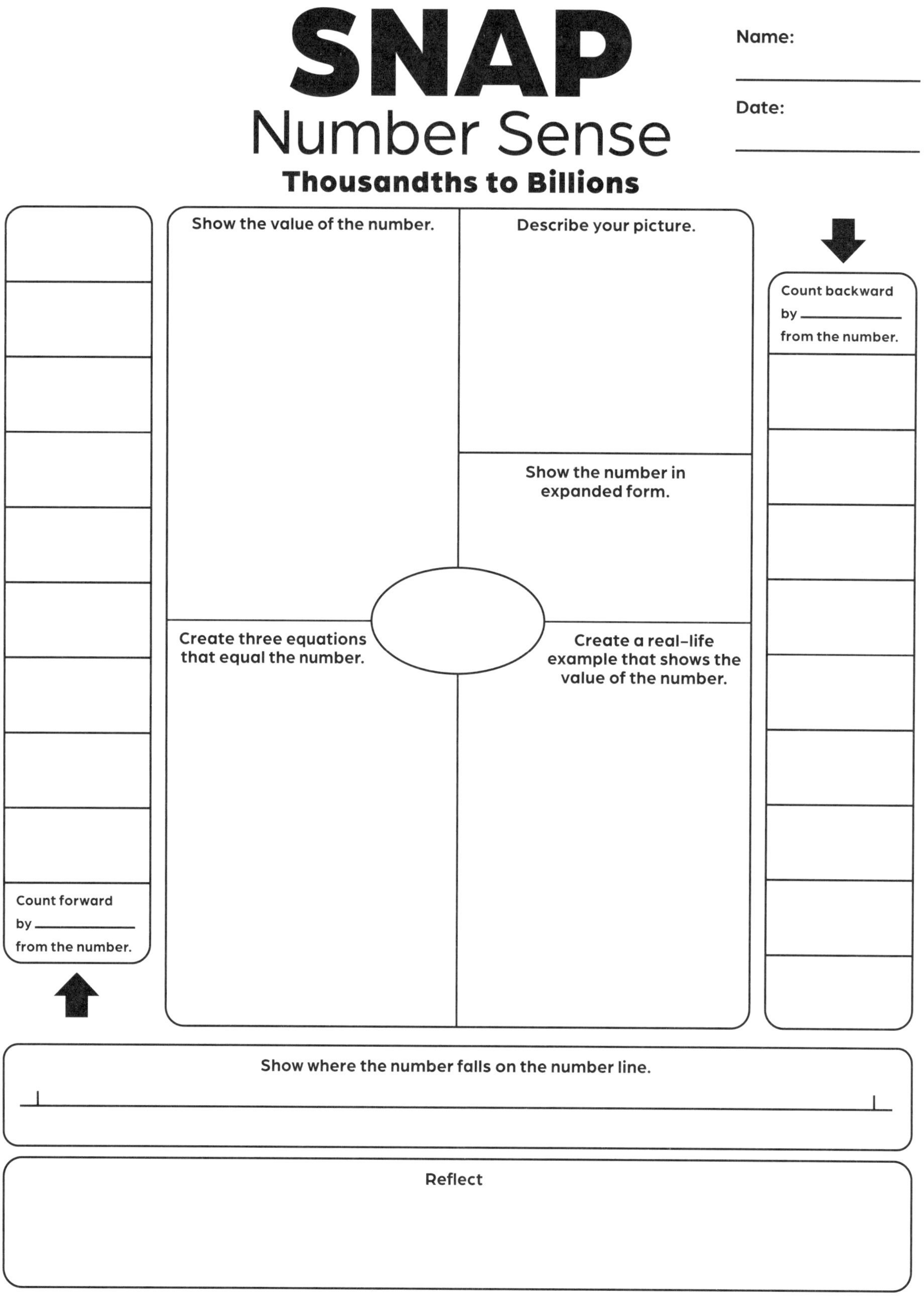

# SNAP
## Number Sense
### Thousandths to Billions

Name:

Date:

**Show the value of the number.**

**Describe your picture.**

Count backward by _____ from the number.

**Show the number in expanded form.**

**Create three equations that equal the number.**

**Create a real-life example that shows the value of the number.**

Count forward by _____ from the number.

**Show where the number falls on the number line.**

**Reflect**

Source: Chilliwack School District (2016). Used with permission.

**FIGURE 2.3:** The SNAP (thousandths to billions).

Visit **go.SolutionTree.com/mathematics** for a free reproducible version of this figure.

> **PRACTITIONER'S PERSPECTIVE**
>
> Chilliwack School District numeracy coach and experienced intermediate teacher Christian Lodders explains that historically, many teachers have looked at mathematics as a series of procedures, rules, and steps that allow you to input numbers and receive an answer. If you can remember and execute those steps, you would get the right answer. If you can't, you would get a wrong answer, leading to a potential origin of the societal distaste seen in mathematics today. For example, in a seven-step procedure, if you did six steps correctly and one incorrectly, your answer was wrong, resulting in no reward or acknowledgment for everything you did well, because the correct answer was the supreme measure of competence.
>
> Christian believes number sense goes far beyond the mathematics class; when students get older and start to ask, "When will I ever use this?" there are hundreds of examples of how they can apply their learned number sense skills (for example, interpreting graphs, estimating volumes, measuring distances, converting units, and understanding ratios).
>
> Like kindergarten students, intermediate students with a strong number sense tend to make fewer mistakes in calculations and reasoning when working in a flexible, creative-thinking-skills-built-in classroom, where number sense skills are essential and generalizable beyond the mathematics class. These students will notice the errors they do make. Because they understand place value and can make reasonable estimates and visualize numbers, students will see when something goes wrong, and they are equipped to take steps to fix it.

Adolescents engage in various activities that require a proficient understanding of number sense (such as shopping, calculating tips at restaurants, and playing video games). Adults rely on number sense to perform these tasks and many others, including tax preparation, calculating mortgages, investing in the stock market, and assessing the impact of inflation on household budgeting. Researchers Robert E. Reys, Barbara J. Reys, Nobuhiko Nohda, and Hideyo Emori (1995) postulate:

> Approximately 80% of mathematical computations in daily life require the mental manipulation of numerical quantities rather than the usage of traditional algorithms. In a mathematically literate society, individuals can think flexibly with numbers, whether mentally calculating the best

> value at the grocery store, estimating the return of money market funds, or checking the reasonableness of a calculator result. (p. 5)

## Classrooms as Safe Spaces

Classroom cultures that provide ethical spaces for teachers to challenge students and for students to challenge others are vital to fostering the ideal conditions for transformational learning. Successful, proactive teachers who create these conditions nurture relationships with their students, create engaging lessons, ensure an inviting classroom environment, and establish routines. In addition, designing safe spaces for students to inquire and take learning risks are key aspects of fostering mathematical competency in students. Researcher and associate professor Sashi Sharma (2015) explains:

> Students' beliefs and attitudes towards risk taking can impact their mathematics learning and performance. However, at present, the risk is not established in the field of mathematics education. The challenge for mathematics teachers in developing their students' risk taking dispositions is to choose appropriate activities and tools that match this concept and the learning needs of the students. (p. 1)

Establishing classroom cultures that promote student learning and allow different pedagogies to flourish for a whole class, small groups, and individually, is a foundational aspect of the SNAP implementation.

## Chapter Summary

This chapter provides an overview of the research regarding number sense foundations, highlighting the important role that number sense plays in mathematics education. Using the voices of classroom practitioners to provide context, we demonstrate the connection between number sense and the SNAP in grades K–8. Finally, we briefly discuss the importance of creating safe and ethical spaces for students to take learning risks in areas such as the SNAP. We acknowledge that teaching and assessing mathematics is very challenging, but we believe the SNAP provides timely and relevant information that addresses specific areas needed to drive learning. Our journey began by addressing these very challenges. In chapter 3 (page 41), we portray how the SNAP corresponds to the five strands of mathematics proficiency set by the National Research Council (2001) from the perspective of grades 2–8 teachers.

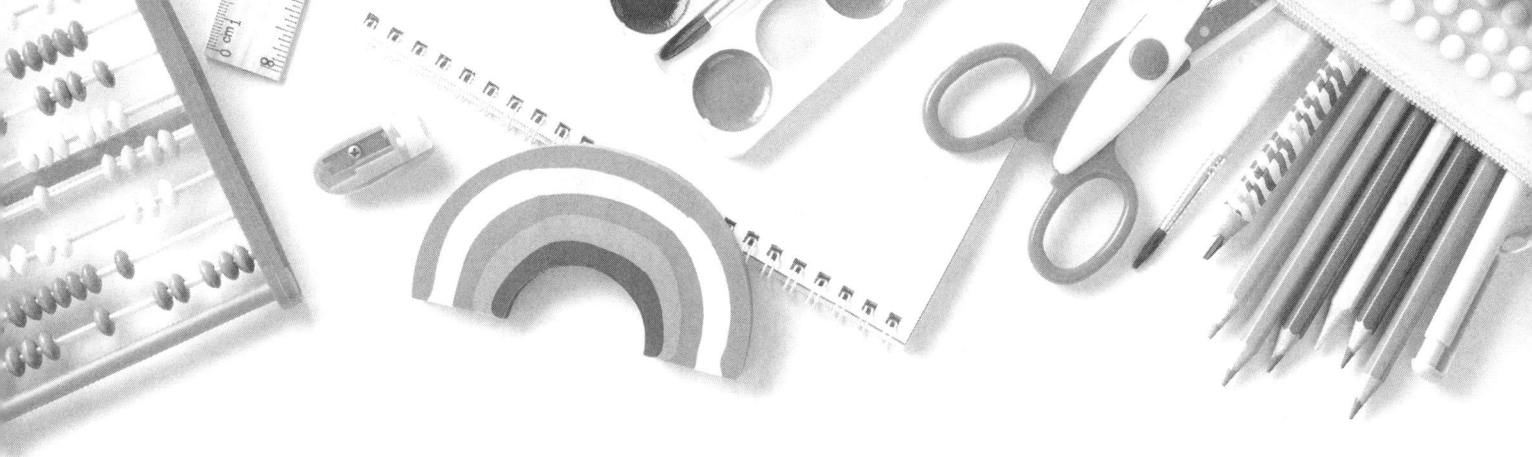

# CHAPTER 3

# Understanding the Five Strands of Mathematical Proficiency

As educators scaffold new learning, breaking it down into manageable chunks for students, they may teach ideas in isolation first, and then bring those ideas together to ensure coherence and connectedness. Therefore, although educators may teach number sense in isolation first, they mainly teach it in an intertwined fashion integrating the five strands of mathematical proficiency (see page 19). Many of the skills in the SNAP demonstrate more than one strand of mathematical proficiency. To see how this works, we categorized the strands to highlight how an educator might introduce and scaffold each individual proficiency (see figure 3.1). It's important to highlight the importance of differentiated instruction as teachers must adjust their instruction, regardless of grade level. In this chapter, we address all five strands of mathematical proficiency (National Research Council, 2001) and explain how the SNAP helps students demonstrate each strand (see figure 3.2, page 42).

| Five Strands of Mathematical Proficiency | The SNAP | |
|---|---|---|
| Conceptual understanding | • Draw to represent the number. | • Write to describe your picture. |
| Procedural fluency | • Write the number in expanded form. | • Count forward and backward from the number. |
| Strategic competence | • Write a real-life example of the number. | • Create three equations that equal the number. |
| Adaptive reasoning | • Show where the number belongs on the number line. | |
| Productive disposition | • Reflect on your thinking. | |

**FIGURE 3.1:** The five strands of mathematical proficiency and the SNAP.

# SNAP
## Number Sense
### 0–10,000

Name: _____

Date: _____

**Show the value of the number.**

**Describe your picture.**

**Count backward by _____ from the number.**

**Show the number in expanded form.**

**Create three equations that equal the number.**

**Create a real-life example that shows the value of the number.**

**Count forward by _____ from the number.**

**Show where the number falls on the number line.**

0 — 10,000

**Reflect**

Source: Chilliwack School District (2016). Used with permission.

**FIGURE 3.2:** The SNAP template.

Visit **go.SolutionTree.com/mathematics** for a free reproducible version of this figure.

## Conceptual Understanding

*Conceptual understanding* refers to comprehending mathematical concepts, operations, and relations. It is the students' understanding of the reasoning behind a task and the ability of students to answer the *why* of mathematics. For example, many teachers have experienced their students asking, "Where will I use this in real life?" There are many answers to this question; ideally, the SNAP assists in helping students see the many connections, usefulness, and wonders of mathematics because it requires students to consider mathematics through the multiple lenses of the five strands of mathematical proficiency.

Teachers can separate the mathematical concepts into individual skills or show the concepts as coherent ideas overflowing with interrelated components. The SNAP provides opportunities for students to practice these skills on their own and see how they interconnect. When learning to read, students learn letters, letter sounds, words, and sentences, eventually building up to paragraphs, stories, and books. The *Common Core State Standards for English Language Arts & Literacy in History/Social Studies, Science, and Technical Subjects* includes skills such as following agreed-on rules for discussions, asking questions to check for understanding, acknowledging new information others express, and initiating and participating in a range of collaborative discussions (NGA & CCSSO, 2010a). In mathematics, students learn numerals, number quantities, times tables, and simple equations, building up to complex algorithms and word problems (NGA & CCSSO, 2010b). Conceptual understanding is essential in both subject areas because it provides a foundation for students to build their understanding as they interact with words and numbers separately and together. Deep levels of comprehension are vital in both language arts and mathematics and are foundational to conceptual understanding. In summary, the National Research Council (2001) explains:

> Students with conceptual understanding know more than isolated facts and methods. They understand why a mathematical idea is important and the kinds of contexts in which it is useful. They have organised their knowledge into a coherent whole, which enables them to learn new ideas by connecting those ideas to what they already know. (p. 115)

The two elements of the SNAP that incorporate conceptual understanding are (1) draw to represent the number and (2) write to describe your picture.

### DRAW TO REPRESENT THE NUMBER

Encouraging students to use drawings to depict the value of a number is a creative and effective way to explore and represent meaning and overall number sense. Drawing helps students express their sense of wonder about mathematics. Creating an image helps students see numbers as more than just digits, but as quantities. Drawing an image of a number encourages students to see numbers in flexible ways. According to speakers and researchers Carole Saundry and Cynthia Nicol (2006), "The ability to make and manipulate a mental image is seen to be key in problem-solving, leading a child to think more flexibly in response to a problem" (p. 58). At an early age, students enjoy representing numbers through drawings. We believe the ability to demonstrate thinking through drawing promotes a personalized approach and leads to equity in classrooms. Using a multimodal approach, students have more opportunities to demonstrate their learning without solely leaning on traditional methods (which rely on written output).

In chapter 4 (page 61), we discuss how beginning mathematicians in kindergarten and grade 1 demonstrate students' thinking using the SNAP. In grade 2, students begin with base ten manipulatives. Learners need to understand what 1s, 10s, and 100s look like, and that they can represent numbers in multiple ways. The concept of place value begins to develop in the early grades. *Number talks* (where students discuss their thinking) are a particularly effective way to instill this concept. Sherry D. Parrish (2011), an experienced classroom educator, university professor, and international speaker with a focus on teaching and learning mathematics through inquiry, defines *number talks* this way:

> *Classroom number talks*, five- to fifteen-minute conversations around purposefully crafted computation problems, are a productive tool that can be incorporated into classroom instruction to combine the essential processes and habits of mind of doing math. During number talks, students are asked to communicate their thinking when presenting and justifying solutions to problems they solve mentally. These exchanges lead to the development of more accurate, efficient, and flexible strategies. (p. 199)

For example, teachers can ask students to explore how many ways they can represent *13*, beginning with 10 and showing how many are left over. Students can explore this activity with partners or in small groups to co-construct their answers.

In grades 2–5, teachers have students begin to draw using base ten blocks as a foundation. Because teachers have already introduced base ten to many students, it is an easy-to-understand strategy to represent numbers. As students become more familiar with drawing numbers, they progress to more complex ways to demonstrate their thinking. For instance, figure 3.3 shows the number 33 composed of three 10 sticks and three 1s—add a 1-block to the figure.

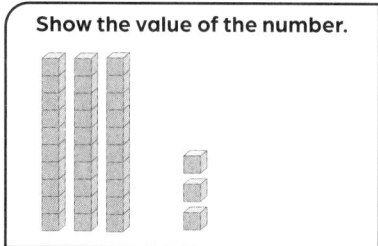

*Source: Chilliwack School District (2016). Used with permission.*
**FIGURE 3.3:** Show the value of the number.

Teachers then encourage students to be creative by using symbols and pictures, including developing unique legends for representing numbers. An example might use three wizards and four broomsticks to describe 34, ensuring the student explains what these visuals represent. Ideally, teachers strive for students to represent numbers using *relative size*. For example, using characters from the *Harry Potter* series by J. K. Rowling, three Hagrids could represent three 10s, and four Harry Potters (as first-year students) could be four 1s. The goal is to engage and ensure students understand that each place value is ten times larger than the previous place value.

Base ten representation for drawing continues to be a popular method for teaching students in grades 6–8. In these grades, teachers introduce even more complex ways of having their students draw to represent numbers, such as having students represent the value of a number using *arrays*, which is another way of putting numbers into groups using rows and columns (see figure 3.4). Another way to challenge students to think visually is using the *multiplication area model*, which requires students divide shapes into areas to calculate their relative size. Researcher and teacher education and administration expert Sarah Smitherman Pratt (2018) reveals that "the measurement model (multiplication area model) for multiplication generates opportunities to conceptualize operations in various ways to elicit more flexibility in understandings" (pp. 17–18). In addition, teachers can incorporate real-life situations, such as *equal sharing*. For example, use apples in baskets, Pokémon cards in stacks, and profits from a lemonade stand to help students visually think about numbers.

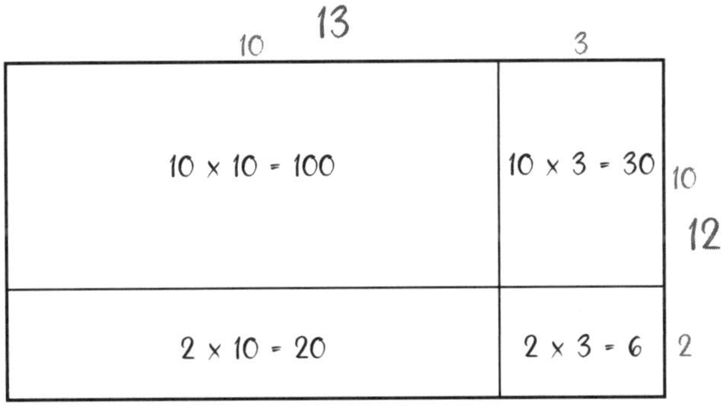

**FIGURE 3.4:** Multiplication area model.

## WRITE TO DESCRIBE YOUR PICTURE

Justifying thinking is an essential skill in any grade, and the ability to answer *why* is a skill of utmost importance (see page 43). In many classrooms across grades K–8, instructors expect students to justify their answers during discussions and in written assignments. Introducing the word *because* to students is a compelling way of promoting justification. An effective sentence stem many teachers use is, "I think _____ is a reasonable answer because _____." The ability to justify has other added benefits, as university researchers and associate professors Kristen N. Bieda and Megan E. Staples (2020) note:

Justification, an essential mathematical practice, is well known for its role in promoting rigor and developing mathematical understanding. Equally powerful is its role as a discursive practice that provides students with *access* to mathematical thinking and reasoning (the mathematics behind the answers) and promotes student *agency* with respect to mathematics. (p. 102)

In kindergarten and grade 1, students orally explain their thinking through discussions with a partner, teacher, or group. In grades 2–8, students not only discuss but teachers also expect them to explain their thinking in writing. When the teacher asks students to describe their picture, the teacher challenges students to move beyond simply stating what they drew and clearly explain their drawing. With larger numbers, students must describe their drawing in reference to the relative scale of numbers as they increase in size. For example, when drawing the number 110, the teacher expects students to note that 100 is ten times larger than 10 (see figure 3.5).

| Describe your picture. | Describe your picture. | Describe your picture. |
|---|---|---|
| I did four squares to represent 400 and I did six lines to represent 60 and one dot to represent 1, so altogether it is 461. | I drew 4 cubes, 2 flats, 8 rods and 6 units. | I used base 10 blocks and a 100 chart (shaded in 62) |
| Grade 3 | Grade 4 | Grade 6 |

*Source: Chilliwack School District (2016). Used with permission.*
**FIGURE 3.5:** Describe your picture.

## Procedural Fluency

*Procedural fluency* refers to the skill of carrying out procedures flexibly, accurately, efficiently, and appropriately. When students are confident and accurate in procedural fluency, they can efficiently and accurately do the mathematics (or perform operations). Fluency connects to *automaticity* (the ability to perform immediately and accurately at a subconscious level), a concept once again in both language arts and mathematics. One way to help young students become fluent in reading is through learning high-frequency *sight words* (such as *the*, *cat*, *with*, and so on) using flash cards. Similarly, a way to help students become fluent in multiplication is having them practice their times tables using flash cards. In both subjects, students must become fluent in basic skills to free up working memory and brain power leading to high levels of comprehension. For example, a cyclist must focus on the journey ahead and not the pedals of the bicycle. Procedural fluency and conceptual understanding directly connect; the National Research Council (2001) reveals, "In the domain of number, procedural fluency is especially needed to support conceptual understanding of place value and the meanings of rational numbers. It also supports the analysis of similarities and differences between methods of calculating" (p. 121). In the SNAP, writing the number in expanded form and counting forward and backward from the number represent procedural fluency.

### WRITE THE NUMBER IN EXPANDED FORM

To promote flexible thinking, students must understand they can break down, take apart, or expand a number into different parts. The idea of *part-part-whole* plays a significant role from kindergarten to grade 2, and leads to the concept of place value when the teachers introduce numbers larger than ten. As coauthors and researchers John F. Cawley, Rene S. Parmar, Lynn M. Lucas-Fusco, Joy D. Kilian, and Teresa E. Foley (2007) reveal, *expanded form* (or expanded notation), or writing a number to show the value of each digit, helps students understand key elements of place value:

> The transposition from pictorial representations is most commonly undertaken by presenting the pictorial form and then moving directly to the "short form." However, many students do not make the transition to the "short form" because they lose the meaning between the . . . hundreds when represented pictorially and when written symbolically. The transposition can be made more understandable with the assistance of expanded notation. (p. 38)

In grade 2, *base ten* is a natural way (ten fingers and ten toes) to have students begin working on expanded form. The expanded form begins with separating groups of 5s and 10s and then transitioning into decomposing or breaking numbers into two smaller parts. Students need lots of practice in the context of place value before working with larger numbers. Once students clearly understand the value of numbers, they can explore groups of ten and hundreds, knowing that the value of a number has many connections to real life (see figure 3.6).

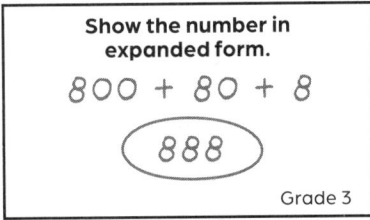 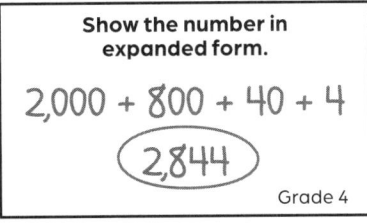 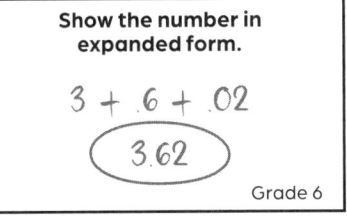

*Source: Chilliwack School District (2016). Used with permission.*
**FIGURE 3.6:** Write the number in expanded form.

Teachers in grades 3–8 believe expanded form is an important concept to teach students; it reinforces place value by separating numbers into meaningful parts. In a discussion on the importance of expanded form, consultant and researcher Nicky Roberts (2019) postulates:

Rather than have children overload their working memory, teachers should encourage children to:

- record intermediate steps (rather than expecting these to be held mentally)
- work flexibly in terms of either from "right to left," or "left to right"
- use structured drawings, together with expanded written methods. (p. 8)

To enhance engagement, some teachers call expanded form *exploded form*, which is another way to explain breaking a number into its relative values or parts and rebuilding a number into its original number. A model using the concept of a numerical water balloon dropping onto the ground with the particles of water (numbers) spraying off in different directions into their corresponding place values could help students. This metaphor helps students understand numbers are made up of pieces (like water droplets) the students can rebuild and move around. Using whiteboards, students breaking numbers into their components can help solidify understanding. Then, students can discuss place value names and explain how they can break all numbers into 1s, 10s, and 100s, with varying family names (that is, *thousands, millions, billions*, and so on). Teachers then take time to break down increasingly complex numbers into their different place values and allow students opportunities to challenge themselves with more complicated numbers (see figure 3.7).

Students typically learn to write numbers in expanded form during their early years and continue to develop this skill as they progress through later grades. The familiar idea of base ten blocks can help reinforce their understanding that numerals consist of digits with different place values. Students must learn to deconstruct large numbers this way, as it is a fundamental aspect of number sense. As students work with increasingly larger numbers, it is important to use manipulatives (like base ten blocks) and tools (such as place value charts and whiteboards) to assist them in comprehending the concept of place value. The SNAP supports students and continues to reinforce this skill in higher grades. The SNAP accomplishes this conceptually by having students draw to represent and explain their drawing, write the number in expanded form, and place the number on a number line. The SNAP supports students and continues to reinforce the skill of representing the value of a digit based on its position in higher grades.

## COUNT FORWARD AND BACKWARD FROM THE NUMBER

Teachers introduce the concept of *skip counting* to students at an early age through rhymes such as, "Two, four, six, eight, who do we appreciate?" Games on the playground (like hopscotch and four-square) expose students to real-life examples of counting forward and backward. The idea of *more* and *less* is an important building block in mathematical thinking. Having the ability to move forward and backward from a number builds student flexibility and confidence and is the building block for addition and subtraction. Counting forward and backward also helps students develop an understanding of relative size and how numbers relate to one another (Van de Walle, Karp, & Bay-Williams, 2013).

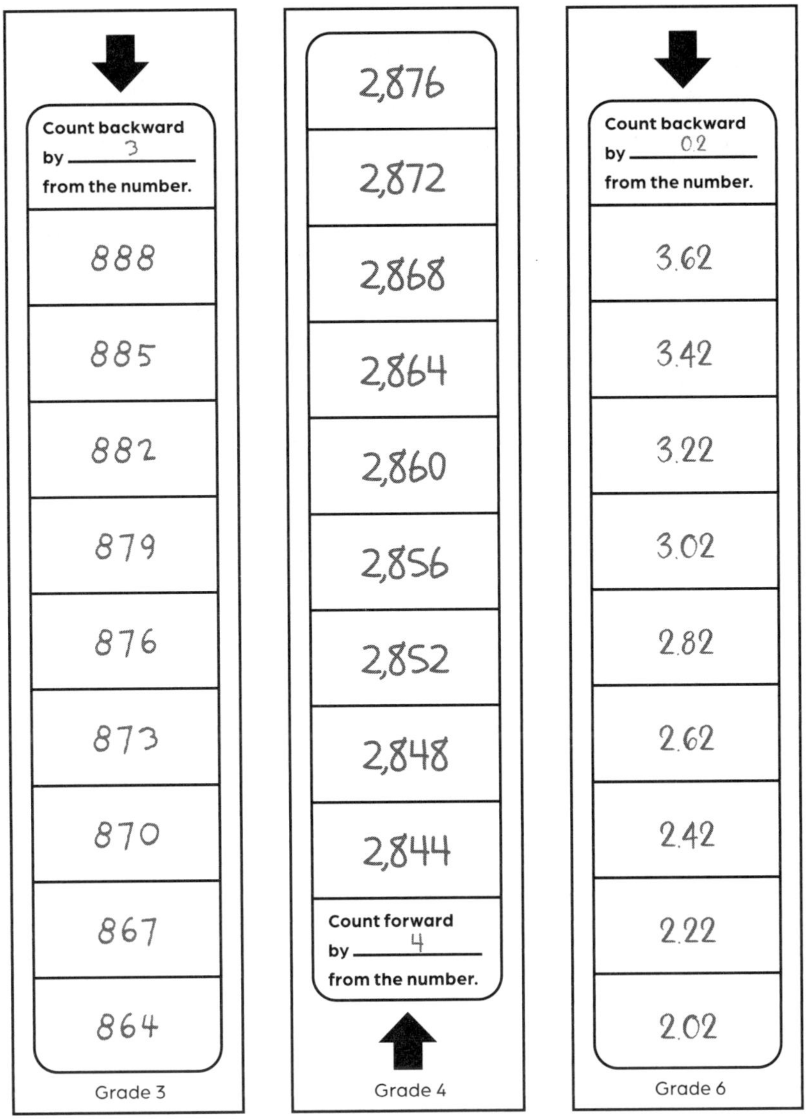

**Grade 3** — Count backward by 3 from the number.
888, 885, 882, 879, 876, 873, 870, 867, 864

**Grade 4** — Count forward by 4 from the number.
2,876, 2,872, 2,868, 2,864, 2,860, 2,856, 2,852, 2,848, 2,844

**Grade 6** — Count backward by 0.2 from the number.
3.62, 3.42, 3.22, 3.02, 2.82, 2.62, 2.42, 2.22, 2.02

*Source: Chilliwack School District (2016). Used with permission.*
**FIGURE 3.7:** Counting forward and backward.

Students enjoy this aspect of the SNAP. By choosing a number (other than 1), students are provided opportunities to work flexibly with numbers, strengthening their procedural fluency.

Playing games by having students count by 2s and 3s is very helpful in teaching this concept in grade 2. For example, using a number line to skip count can assist students in seeing patterns and relationships. Other visuals (such as the image of an elevator or thermometer) can support students in seeing how numbers can get larger or smaller in real life. The SNAP requires students to complete skip counting vertically, so it's also important to have students practice skip counting horizontally as well, which is similar to using a number line (see figure 3.7).

While some intermediate teachers may find skip counting straightforward for students, others may notice this area is a challenge for students; those teachers must work with struggling learners to reinforce their understanding of increasing and decreasing numbers. It can be challenging for students who have yet to solidify their understanding when a number rolls over into the next place value. For example, 37 may become 311 instead of 41 (37 plus 4 more), or 399 + 2 could become 3,911 instead of 401. To help students understand this concept, teachers often break down the skill into small numbers that roll over into the 10s place value, using manipulatives to reinforce student understanding. As students become more comfortable with the concept, teachers gradually introduce more challenging numbers and gradually remove manipulatives to reinforce automaticity.

Like in earlier grades, teachers in grades 6–8 encourage students to practice skip counting from any number by 2s, 3s, 4s, 5s, and 10s to begin. Teachers can use a hundreds chart (see figure 3.8) and other charts to build students' pattern-searching skills and then have the students explain their thinking. For example, if a student starts at 117 and skip count by 5s (that is, 117, 122, 127, 132, 137, and so on), what pattern will emerge in the 1s place? The digit in this place will alternate between 2s and 7s. If a student starts at 117 and skip count by 4s (that is, 117, 121, 125, 129, 133, 137, 141, 145, 149, 153, 157), what pattern will emerge in the 1s place? The pattern core is 7, 1, 5, 9, and 3. The sample hundred chart in figure 3.8 is a useful tool to aid students in skip counting.

| 1 | 2 | 3 | 4 | 5 | 6 | 7 | 8 | 9 | 10 |
|---|---|---|---|---|---|---|---|---|---|
| 11 | 12 | 13 | 14 | 15 | 16 | 17 | 18 | 19 | 20 |
| 21 | 22 | 23 | 24 | 25 | 26 | 27 | 28 | 29 | 30 |
| 31 | 32 | 33 | 34 | 35 | 36 | 37 | 38 | 39 | 40 |
| 41 | 42 | 43 | 44 | 45 | 46 | 47 | 48 | 49 | 50 |
| 51 | 52 | 53 | 54 | 55 | 56 | 57 | 58 | 59 | 60 |
| 61 | 62 | 63 | 64 | 65 | 66 | 67 | 68 | 69 | 70 |
| 71 | 72 | 73 | 74 | 75 | 76 | 77 | 78 | 79 | 80 |
| 81 | 82 | 83 | 84 | 85 | 86 | 87 | 88 | 89 | 90 |
| 91 | 92 | 93 | 94 | 95 | 96 | 97 | 98 | 99 | 100 |

**FIGURE 3.8:** Hundreds chart.

## Strategic Competence

*Strategic competence* refers to the ability to formulate, represent, and solve mathematical problems. If there is one area where students and adults struggle, it is problem solving. Reading can be a hurdle here, and trying to understand what a problem is asking can be very challenging, let alone solving the problem accurately once the student has determined what the question is asking. The SNAP takes a different approach by beginning with the end in mind and asking students to create their own word problems. Knowing that problem solving is a challenging concept for students, the SNAP provides students an opportunity to create a real-life example of a problem together with being able to solve it. The National Research Council (2001) reinforces this notion:

> Although in school, students are often presented with clearly specified problems to solve, outside of school, they encounter situations in which part of the difficulty is to figure out exactly what the problem is. Then they need to formulate the problem so that they can use mathematics to solve it. Consequently, they are likely to need experience and practice in problem formulating as well as in problem solving. (p. 124)

The SNAP addresses strategic competency when students complete these two tasks: (1) write a real-life example of the number and (2) create three equations that equal the number.

### WRITE A REAL-LIFE EXAMPLE OF THE NUMBER

The ability to take a concept and apply it in real life is a fundamental objective of education. The turning point in any lesson should include allowing students to transform their learning and use it within their own context. In real life, students observe numbers in supermarkets, on phones, at gas stations, on sports jerseys, and so on. Knowing there is a difference between a numeral and the idea that the numeral contains a quantity (or an amount) can go unnoticed, which requires students to focus on objects differently. Researchers M. Maghfirah and Ali Mahmudi (2018) reinforce the important of connecting mathematics to real life when they reveal:

> Not all students in the future will become a mathematician, but every student who learns math is expected to apply their knowledge that they get in everyday life. The knowledge is not merely vis-à-vis the use of standard algorithms such as the use of complex formulas and written calculations, but intuitions built on meaningful learning experiences. (p. 2)

The opportunities to see numbers in real life are abundant and exciting. We encourage teachers to take advantage of this section of the SNAP to stretch students' thinking.

Grade 2 teachers use exploring in and outside the classroom to strengthen this concept. When teachers invite students to examine the mathematical world around them, the students are delighted with what they discover. Taking students outdoors and conducting number walks is incredibly effective, and students are intrigued and motivated to discover numbers in nature. *Measurement* is a practical way to teach students about numbers in real life. For example, students can cook using measuring cups, measure their height using yardsticks, or step on scales to record their weight. Students also enjoy using real-life manipulatives such as pennies, nickels, and dimes to understand different practical mathematical concepts. These manipulatives also lend themselves to teachers introducing cross-curricular concepts to show how mathematics transcends the time the teachers devote to it.

In the intermediate grades, the objective is to help students differentiate between the real-life representation (meaningful quantity) of a number versus a symbolic use (when they have previously encountered it). For instance, writing the statement, "Wayne Gretzky's jersey number is 99" does not imply a proper understanding of the value of the number 99. Teachers provide examples of how a number relates to a real-life quantity, and encourage students to share their own examples to encourage further creative thinking for future real-life scenarios. For example, "I received 34 chocolate bars while trick-or-treating, but my dad ate most of them!" As numbers increase in value, it becomes increasingly important to share these examples, as real-life applications may become more challenging for students to grasp. For example, the number one million can be difficult for students to express in real life. The following vignette shows how one student works with large numbers.

> **Chase is a 10-year-old student who is really connected with creating the real-life examples—especially with large numbers. He is an aviation fan and knows how many seats there are on a Boeing 777. When working with large numbers, Chase can recite exactly how many jets there are, including how many seats are left over (see figure 3.9). For example, if the number in the middle of the SNAP is 10,000, Chase would explain in his real-life example that 10,000 is the number of seats in 47 Boeing 777s, with approximately 17 seats left over. (We expect him to become a pilot one day!)**

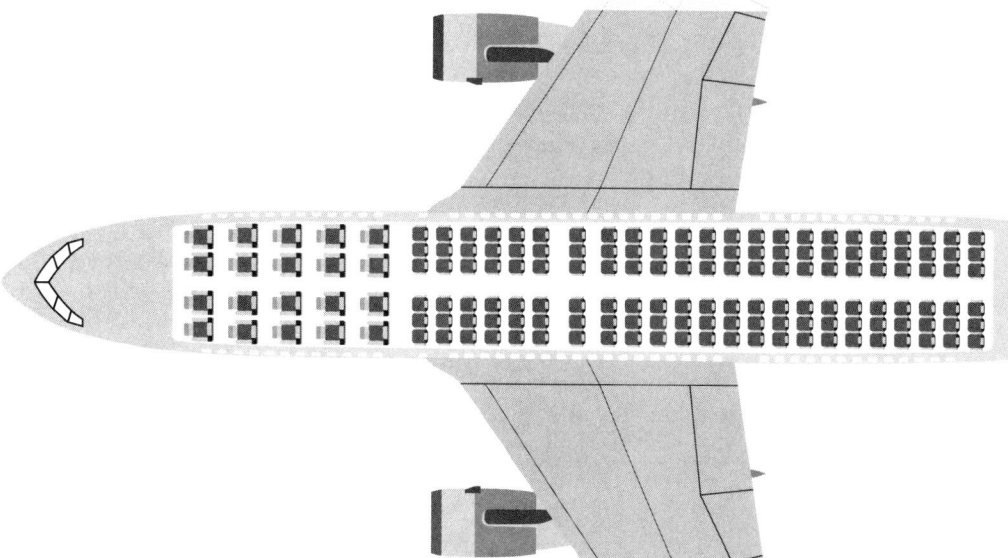

**FIGURE 3.9:** Connecting with a real-life example of a plane.

Students in grades 6–8 who need help with creating word problems tend to have the fewest memorable life experiences from which to draw examples. To support these students, as well as *all* students, we suggest the following strategies.

- Take the class on field trips and point out the mathematics in different situations.
- Use nature as a source of numeracy experiences.
- Point out the mathematics in students' interests (for example, Minecraft, YouTube channel analytics, sports statistics, and so on).
- Point out the mathematics in class read-aloud stories.
- Use online resources like Estimation 180 (https://estimation180.com) to help students build a conceptual understanding of larger amounts in real situations.
- When exploring five- and six-digit numbers, encourage the use of money, populations, and distances. (For example, a family house in Spokane, Washington, might cost $845,999, the circumference of the Earth at the equator is 24,901 miles, and so on.)

## CREATE THREE EQUATIONS THAT EQUAL THE NUMBER

Asking students to create three equations that equal a number is another way for them to become more proficient with number sense. It is common for students to encounter equations during mathematics classes, but what is unique is seeing equations as part of a broader context. The SNAP does this by having students see many interrelationships and allowing them to stretch their thinking in new ways. Creating equations is an important aspect of number sense, as University of Cyprus mathematics educators Marios Pittalis, Demetra Pitta-Pantazi, and Constantinos Christou (2015) explain:

> The inclusion of algebraic arithmetic as a factor of number sense results in reconceptualising the conception of number sense. The proposed nature of number sense defines a more dynamic and flexible construct that could facilitate students' advancements and transition to a more abstract and relational system of thinking. (p. 447)

When beginning in this section of the SNAP, students start with simple equations, and as they become more adept with simple equations, they progress to more complex questions (figure 3.10, page 54). Teacher feedback and dialogue can help motivate students to push themselves to create complexity and relevance in their equations.

| Create three equations that equal the number. | Create three equations that equal the number. | Create three equations that equal the number. |
|---|---|---|
| 888 × 1 = 888 | 3,000 + 1,286 = 4,286 | 1.81 × 2 = 3.62 |
| 700 + 188 = 888 | 6,000 − 2,000 + 286 = 4,286 | 7.24 ÷ 2 = 3.62 |
| 977 − 89 = 888 | (2 × 2,000) + 286 = 4,286 | 4.77 − 1.15 = 3.62 |
| Grade 3 | Grade 4 | Grade 6 |

*Source: Chilliwack School District (2016). Used with permission.*
**FIGURE 3.10:** Create three equations that equal the number.

When teachers use base ten in grade 2, they must introduce addition and subtraction slowly, so students can gain confidence. As in all effective learning, connecting a new idea to what students have already encountered is advantageous. Counting backward and forward on a number line is very useful in having students create various equations that equal a number. Using rekenreks is another excellent way of reinforcing the idea of creating simple addition and subtraction equations.

Creating equations is exciting for grades 3–5 students *and* teachers. Students must have a good grasp of number sense and operational skills to excel at creating equations. Initially, students may feel overwhelmed, so begin with simple addition. We recommend creating straightforward equations such as 33 + 1 = 34. Teachers often dedicate many lessons to help students improve their skills in this area. For example, by the end of the year, if grade 4 students can solve grade-level equations that include one addition, one subtraction, and one multiplication question, then the SNAP has been successful.

Lessons introducing and reinforcing addition, subtraction, multiplication, and division concepts using concrete pictorials to abstract progressions (that is, not just standard algorithms) are effective in grades 6–8. Regular fluency practice with age- and readiness-appropriate operations is important, and there are various games and activities to promote this learning (see appendix B, page 129). Encourage students to build the most complex equations they can while still maintaining equality and arriving at the correct answer. This section of the SNAP is an opportunity for students to show off what they can do with numbers. For example:

- 45,597 × 3 − 2 = 136,789 (appropriate for some)
- 135,999 + 790 = 136,789 (appropriate for some)
- 136,788 + 1 = 136,789 or 136,789 × 1 = 136,789 (inappropriate)
- 45,587 (round to 50,000) × 3 − 2 = 149,998 (appropriate for students who can take a complex number and simplify it for elegance and brevity)

## Adaptive Reasoning

*Adaptive reasoning* is a process that connects to students' capacity for logical thought, reflection, explanation, and justification. It is an area where students work through mathematical tasks and discuss their reasoning with a partner, in a group, or with the whole class. In traditional classrooms, teachers give students a mathematical task and expect students to complete it in relative silence. In a student-centered classroom, students and teachers explore mathematics concepts, encourage discussion, and promote group work. Utah State University mathematics professor Janice Bodily (2012) explains in her master's thesis that in student-centered classrooms:

> Students feel empowered to share their insights, ideas, and questions about the mathematics being presented. All have the opportunity at one time or another to present and defend their mathematical ideas. Students are prepared to answer questions about their theories. The classroom no longer has the same structure as a traditional classroom where the teacher solely directs the knowledge and ideas that are presented (i.e., sage on the stage). (p. 7)

Instead of simply memorizing and following procedures, teachers expect students to explain their thinking and reasoning in an environment conducive to critical thought. The National Research Council (2001) notes:

> In mathematics, adaptive reasoning is the glue that holds everything together, the lodestar that guides learning. One uses it to navigate through the many facts, procedures, concepts, and solution methods and to see that they all fit together in some way, that they make sense. In mathematics, deductive reasoning is used to settle disputes and disagreements. Answers are right because they follow from some agreed-upon assumptions through a series of logical steps. (p. 129)

In the SNAP, teachers assess students' adaptive reasoning when students can show where the number belongs on the number line.

At first glance, a number line can appear simplistic; however, teachers can glean a great deal of information through teaching adaptive reasoning via formative, short-cycle, and ongoing assessment. Beginning with partitioning, students quickly understand the importance of equal distances and measurement. The visual aspect of the number line allows students to identify a number in time and space relative to other numbers. The arrows on a number line present students with two different options: to the left is less than, and to the right is more than. It is a strategy accessible for students of any age and has applications beyond its use. As coauthors Dawn M. Woods, Leanne Keterline Geller, and Deni Basaraba (2017) posit:

> Moreover, systematically incorporating precise visual representations like the number line into mathematics instruction as early as

> kindergarten and continuing across the grades will not only support the development of students' understanding of increasingly complex mathematics concepts (e.g., proficiency with operations and algebraic thinking) but also equip students with a mathematical tool that can support their thinking over time. (p. 7)

Because this task has multiple access points, number line work is an activity that lends itself exceptionally well to differentiation. Teachers can determine the starting numbers, the maximum and minimum values, and the range of numbers based on each student's current understanding and ability level.

In grade 2, teachers can create a number line that students can always refer to in the classroom. Activities such as giving students numbers on sticky notes and having them place the notes in order on the board can not only be an effective way to promote number sense but also incorporate movement into a numeracy lesson. *Clothesline mathematics activities*, where students put different numbers in order on a clothesline, are also helpful in getting students to understand where numbers belong on a number line.

Grades 6–8 teachers notice that when they initially ask students to place a number (natural, decimal, fraction, or integer) on a number line, they may need help understanding where it should go. To avoid having students randomly place the number, we suggest creating a checklist for students that includes the following.

- Select five anchor points with the starting points included.
- Use the center value and then decide on which side of the center the number belongs.

The usable side can then be divided into logical portions (10s, 100s, 1,000s, and so on). This approach may result in more than five anchor points, but it will help students understand where the number generally belongs.

Starting with a number line that ends at 10 and then progresses to 100, 1,000, and 10,000 is beneficial. With practice, students gain confidence and can work with more complex numbers.

Learning can (and should) happen everywhere! For example, one grade 3 teacher wanted her students to explore mathematics in a variety of locations, so she took her classroom outside and had students create their own number lines using chalk. She provided the students a number and then asked students each to place the number on their own number line. The students enjoyed learning about number lines on the playground, and the level of engagement was phenomenal, as was the student learning.

Again, in grades 6–8, we encourage teachers to display a number line prominently in the classroom. Teachers at these grade levels also use clothesline mathematics activities to provide concrete or visual opportunities for students to see and manipulate values on a number line (see figure 3.11).

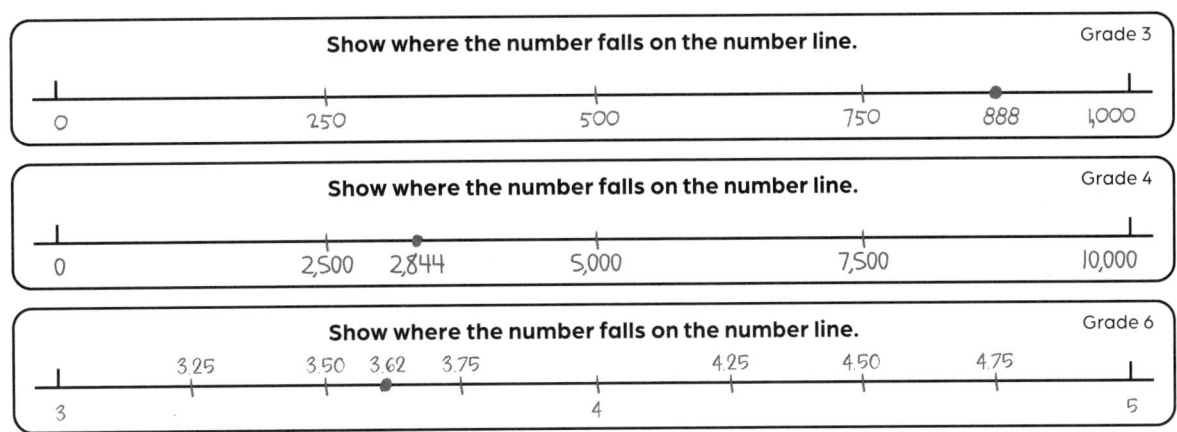

Source: Chilliwack School District (2016). Used with permission.
**FIGURE 3.11:** Values on a number line.

Teachers can challenge students to simultaneously compare various numbers and values in grades 6–8. Rekenreks also continue to be valuable tools to help students learn this concept.

## Productive Disposition

*Productive disposition* refers to the habitual inclination to see mathematics as sensible, useful, and worthwhile, coupled with a belief in diligence and self-efficacy (National Research Council, 2001). *Self-efficacy* (that is, the belief one is competent in mathematics) is an area lacking in many students and adults alike. One of the main purposes of the SNAP is to instill in students a sense of wonder about mathematics. When people are curious about a topic, they are more apt to engage with it. Because some people avoid mathematics, we purposely designed the SNAP to counteract this through a graphical organization. The SNAP resembles a puzzle to solve or a game to play, as an aspect of puzzles and games is the idea of sticking with it (or being diligent) until it is finished. It is also important for students to have numerous opportunities to create meaning and understand the meaning behind mathematics. The National Research Council (2001) writes:

> If students are to develop conceptual understanding, procedural fluency, strategic competence, and adaptive reasoning abilities, they must believe that mathematics is understandable, not arbitrary; that, with diligent effort, it can be learned and used; and that they are capable of figuring it out. Developing a productive disposition requires frequent opportunities to make sense of mathematics, to recognize the benefits of perseverance, and to experience the rewards of sense making in mathematics. (p. 131)

Teachers assess the students' productive disposition competency in the Reflect section of the SNAP (see figure 3.12, page 58).

| Easy | Reflect | Hard | Grade 5 |
| --- | --- | --- | --- |
| Drawing to represent the number was easy because I've been doing it for so long I just remember it. | | Writing a real-life example was hard because when you have such a big number it's hard to relate to real-life things. | |

Source: Chilliwack School District (2016). Used with permission.
**FIGURE 3.12:** Student reflections in the SNAP.

As educators, it is important to teach students about *metacognition* (or thinking about your thinking). Metacognition is a lifelong skill educators must teach and reinforce in every grade. This learning can occur through reflective exercises, such as asking students "What did you feel good about?" or "Where did you gain confidence during the lesson?" Setting expectations at the beginning of a lesson or activity with *I can* statements and regularly ending lessons or activities with questions will promote reflective thinking. Teaching students how to reflect during the school day, especially at the end of each lesson, is vital to seal in the learning—much like closing a plastic bag to preserve the freshness of its contents. Some examples of reflective questions include the following.

- How did you become a more confident mathematician today?
- What part of this mathematics concept do you feel confident with?
- What was your greatest accomplishment in today's mathematics lesson?

When students set a goal and then reflect on whether they achieved the goal and what they will do next time, it promotes deeper thinking in all areas. Researchers Hui Fang Huang Su, Frederick A. Ricci, and Mamikon Mnatsakanian (2016) discuss the importance of metacognition, stating:

> When teaching mathematics options for solving problems or during computations, teachers can assist students by expanding those math reasoning skills associated with advanced mathematics, which require a higher level of thinking, critical thinking or thinking about thinking (often referred to as metacognition). (p. 191)

Students often enjoy sharing their struggles, so using the Reflection section of the SNAP helps them articulate their frustrations and understand that others may share the same challenges. By sharing these reflections aloud, students can develop learning goals for the next time. Unlike in kindergarten to grade 2, in grades 3–8, student reflections are more direct and refined, and teachers expect students to share their thinking with partners, in large groups, and in detailed written form. Teachers expect more justification of thinking and volume of writing. The skill of reflecting on thinking plays a pivotal role in students becoming independent rather than passive thinkers, and the SNAP allows students to hone their skills in this area. Teachers can also emphasize the positive nature of productive struggle by keeping track of success and near misses. Sometimes, students find the joy of mathematics in getting multiple right answers through novel and unconventional ways. However, through dialogue it may become clear that the method is a one-off solution and not a practical mathematics strategy.

## Chapter Summary

As each skill area is represented under the different headings in the SNAP and also in the five strands of mathematical proficiencies, it is important to revisit the idea that the SNAP is intended to highlight the many connections in number sense in accessible ways that promote student curiosity. Although educators can examine the strands of conceptual understanding, procedural fluency, strategic competence, adaptive reasoning, and productive disposition on their own (much like the different components in the SNAP), the value is seeing how all the strands interweave into a common thread. In chapter 4 (page 61), we discuss the SNAP adaptations for kindergarten and grade 1.

# CHAPTER 4

# Implementing the SNAP With Beginning Mathematicians

There is no experience quite like stepping into a kindergarten or a grade 1 classroom—so much happens simultaneously! At mathematics centers, young students explore and play with various objects, building shapes and patterns using the full extent of their imaginations. Students in these grades are genuinely surprised when they discover ten paper clips are the same amount as ten pencils despite the noticeable visual differences. At this age, students learn how to subitize a number or identify the quantity of a pattern without counting its components. Additionally, they begin to explore one-to-one correspondence and the concepts of decomposing and recomposing numbers. Through play, teachers challenge early learners to understand various number relationships, and students begin to deepen their understanding of number sense. As students progress, they continue to work on foundational number sense skills by focusing on reasoning, analyzing, understanding, solving, communicating, representing, connecting, and reflecting. Throughout this chapter, we discuss the origins of the SNAP in both kindergarten and grade 1. We also provide ideas for how to implement the SNAP at these grade levels.

## Implementing the SNAP in Kindergarten and Grade 1

The idea behind the SNAPs for kindergarten and grade 1 is to scaffold students for what they will encounter in grade 2. Therefore, three teachers from Strathcona Elementary in the Chilliwack School District took the previously designed and implemented grade 2 SNAP and incorporated elements of it into new versions for kindergarten and grade 1 (see figure 4.1). The primary team then met with groups of district kindergarten and grade 1 teachers, which led to creating a K–1 SNAP pilot. Within months, many neighboring school districts also took an interest in implementing the SNAPs for kindergarten and grade 1.

The primary team designed the new kindergarten and grade 1 SNAPs for use both as a whole class and one-to-one assessment with students. Therefore, teachers can use the grade-appropriate templates in various ways with their whole class at the beginning of the year. For example, some teachers record student responses on the SNAP, while others place the SNAP under a document camera or recreate it on a whiteboard while leading the students in a class number talk. The teacher chooses a number and guides the class through how to represent the number in an "I do, we do, you do" fashion, which gradually releases the responsibility of doing the task to the students. Students then use manipulatives or whiteboards to create the number. We recommend this format each time the teacher introduces a new concept. Despite the time demand, teachers can glean a great deal of information about students' thinking using the SNAP in a one-to-one format. With manipulatives, teachers give students a number, and students demonstrate their understanding by performing each set of tasks. Completing the SNAP in this way provides teachers with a window into each student's thinking. Along with direct classroom instruction, teachers encourage students to understand that mathematics is all around them, and they authentically integrate it into various aspects of their day. This could take the shape of counting repetitions together during a movement break, or a cooking class where teachers ask students to stir a mixture ten times.

The SNAP has multiple uses *for* learning. Therefore, teachers must explicitly teach every aspect of the SNAP. Working as a whole class, in small groups, or individually helps clarify whether a student is emerging, developing, applying, or proficient in any given task. Likewise, when conferencing with students, the teacher can notice in real time how a student is thinking and any gaps in understanding. Tracking progress through formal and informal observations, together with summary sheets, helps reveal themes that lead to areas requiring further instruction.

In September, the teacher can introduce a concept through a whole-class conversation and then invite students to participate in an activity or game to check for understanding. (See appendix B, page 129, for suggested activities and games.) Teachers can observe whether students are counting in sequence and then zoom in on students, especially those students who struggle. Observation should occur naturally—without putting students under a microscope or making them feel singled out. Teachers can also pair students with partners, with one partner taking on the role of a coach and saying things like, "I like the way you did this" or "Please tell me more about that." Teachers can build practice into the school day and observe during carpet time, lineups, or centers time. Teachers can also conduct miniconferences with students who need help understanding a given concept throughout the school day.

## Using the SNAP in Kindergarten

The following sections provide instructions for guiding students through the six sections of the kindergarten number sense SNAP (see figure 4.1).

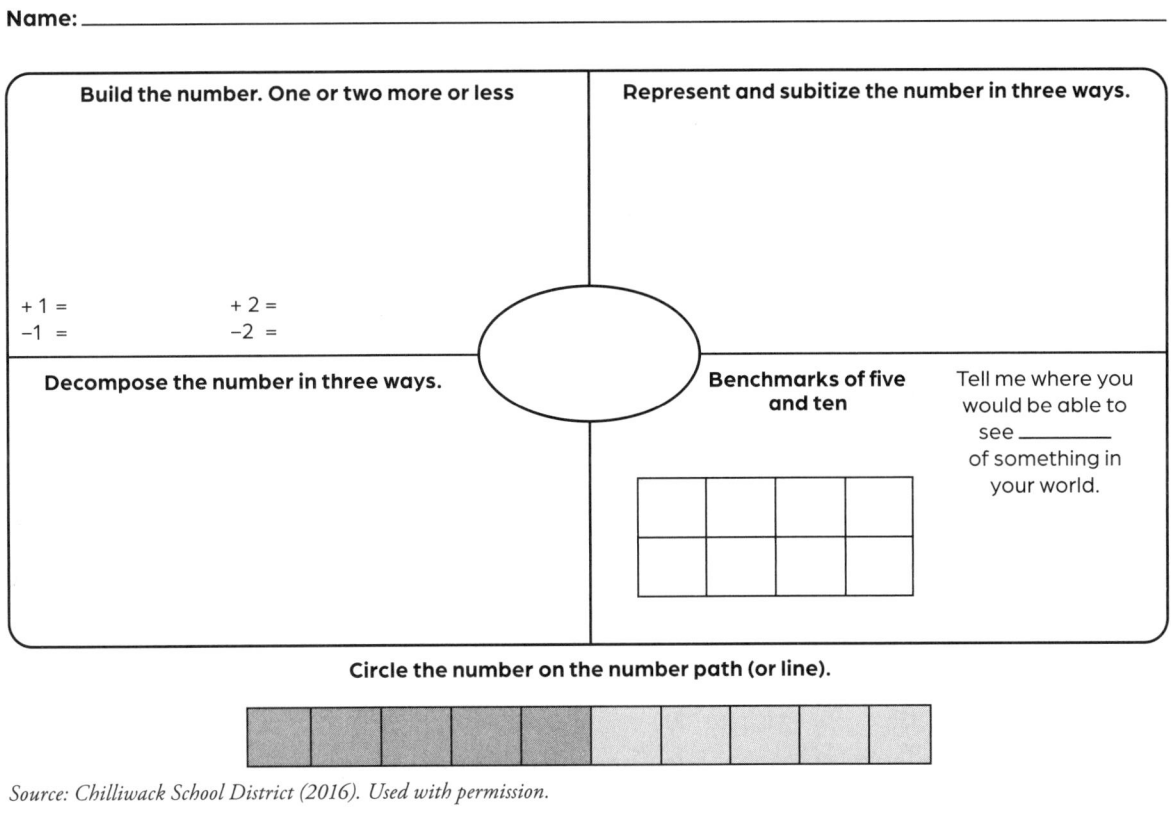

*Source: Chilliwack School District (2016). Used with permission.*

**FIGURE 4.1:** Kindergarten number sense SNAP.

*Visit go.SolutionTree.com/mathematics for a free reproducible version of this figure.*

### BUILD THE IDEAS OF FIVE AND TEN AND ONE MORE OR LESS

Foundational concepts in kindergarten are the ideas of *five and ten* and *one more or less*. Teacher and preservice teachers consultant Janice Novakowski (2007) explains, "Five is an essential benchmark number for young students, and a strong understanding of five will contribute to their understanding of ten, another significant benchmark number in our number system" (p. 1). One of the best ways to introduce students to this concept is by using *ten frames* (two by five rectangular frames students can fill with counters). Using ten frames, students begin to see that five and ten consist of different parts and can group those parts of these numbers in various ways. Using ten frames, students can also see the relationships between more and less. When introducing the idea of building numbers, it is important to begin adding before subtracting because, initially, most students understand the concept of *more* better than *less*. Number talks are another great way to introduce the concept of more or less (for example, the teacher asks, "Let's begin with four. What is one more?" and "Let's begin with four. What is one less?"

### REPRESENT AND SUBITIZE THE NUMBER IN THREE WAYS

Knowing how to *subitize* (or to perceive a number quickly without counting) is an important skill for primary students to master. Coauthors and researchers Cathy Yun and colleagues (2011) submit, "Subitizing ability is believed to underlie the development of fundamental mathematics skills in early childhood and to support mathematics achievement" (p. 1). To reinforce this concept, students draw using a ten frame, dice, tallies, fingers, and alphabet letters. Representations using dots can be helpful; the teacher can ask, "Show me where there are four." If students need to count the dots, they cannot subitize yet, and require further instruction.

### DECOMPOSE THE NUMBER IN THREE WAYS

*Part-part-whole* (or putting two parts together to make a whole or breaking a whole apart into parts) is fundamental to students' understanding of number sense. When students understand they can break down a number in multiple ways, they have acquired the groundwork for advanced mathematical reasoning. Understanding part-part-whole relations connects to students' understanding of numbers and operations (Vlassis et al., 2023). An effective strategy to help students understand is using *decomposing bags*; students can shift objects from one side of the bag to the other, thus having a tangible representation of the notion they can group the numbers in various ways. When working with decomposing bags, students work individually, in pairs, or in small groups, and take apart a number, such as 4, to demonstrate the two parts inside the bag (see figure 4.2).

**FIGURE 4.2:** Using a decomposing bag.

## BENCHMARKS OF FIVE AND TEN

As we state earlier (see page 63), the concept of ten is fundamental in kindergarten. The ability to work in base ten sets the stage for students to be able to think mathematically for years to come. One way to teach benchmarks of five and ten is to have teachers use tallies, money, and *Friends of Ten* (or numbers that add up to 10) and ask the question, "How do we make ten?" Teachers typically begin by introducing the concept of counting to five using fingers and then progress to counting to ten using all fingers. During calendar time, teachers also integrate the days of the week and teach students to group objects into bundles of ten. Finding patterns of five and ten in nature is another excellent way to explain benchmarks of five and ten. Books such as *1, 2, 3 Salish Sea: A Pacific Northwest Counting Book* by Nikki McClure (2021), *Anno's Counting Book* by Mitsumasa Anno (1977), and *Counting on Snow* by Maxwell Newhouse (2017) can also assist students in understanding benchmarks of five and ten.

## TALK ABOUT WHERE MATHEMATICS EXISTS IN THE WORLD

By including real-life examples of numbers early, young students can see that mathematics is all around them. In their position statement, the NCTM (2014) reveals that students should be able to confidently use their mathematics skills to explain applications and analyze situations that arise in the real world. Teachers can promote real-world understanding of mathematics by taking students outdoors and posing questions such as, "How many swings are there on the playground?" or "How many steps are there leading into the school?" While outside, teachers can also ask students to gather five stones, sticks, or leaves or, when back inside the classroom, gather five crayons, pieces of paper, or paper clips. All of these activities promote the sense that students can find mathematics everywhere.

## CIRCLE THE NUMBER ON A NUMBER LINE

Number lines assist students with their ability to see numbers in real life. The ability to place a number on a number line (or path) is important for students' development of overall mathematical understanding (Booth & Siegler, 2008; Geary, Hoard, Byrd-Craven, Nugent, & Numtee, 2007; Schneider et al., 2018). Students enjoy interacting with number lines, which reinforce the idea of benchmarks of ten. For instruction, teachers should start with a number line of ten and then ask the students, "Show me 3 by coloring in three squares." Students can do this activity as a whole class, and students particularly enjoy it when teachers show *non-examples* (for example, coloring in five squares, as the non-example in figure 4.3 shows). Students will quickly point out the error, and then teachers ask why it is incorrect. This strategy helps to build student confidence and reinforces the playful potential of numbers.

**FIGURE 4.3:** Non-example of a number line for kindergartners.

## Using the SNAP in Grade 1

The following sections provide instructions for guiding students through the seven sections of the grade 1 number sense SNAP (see figure 4.4).

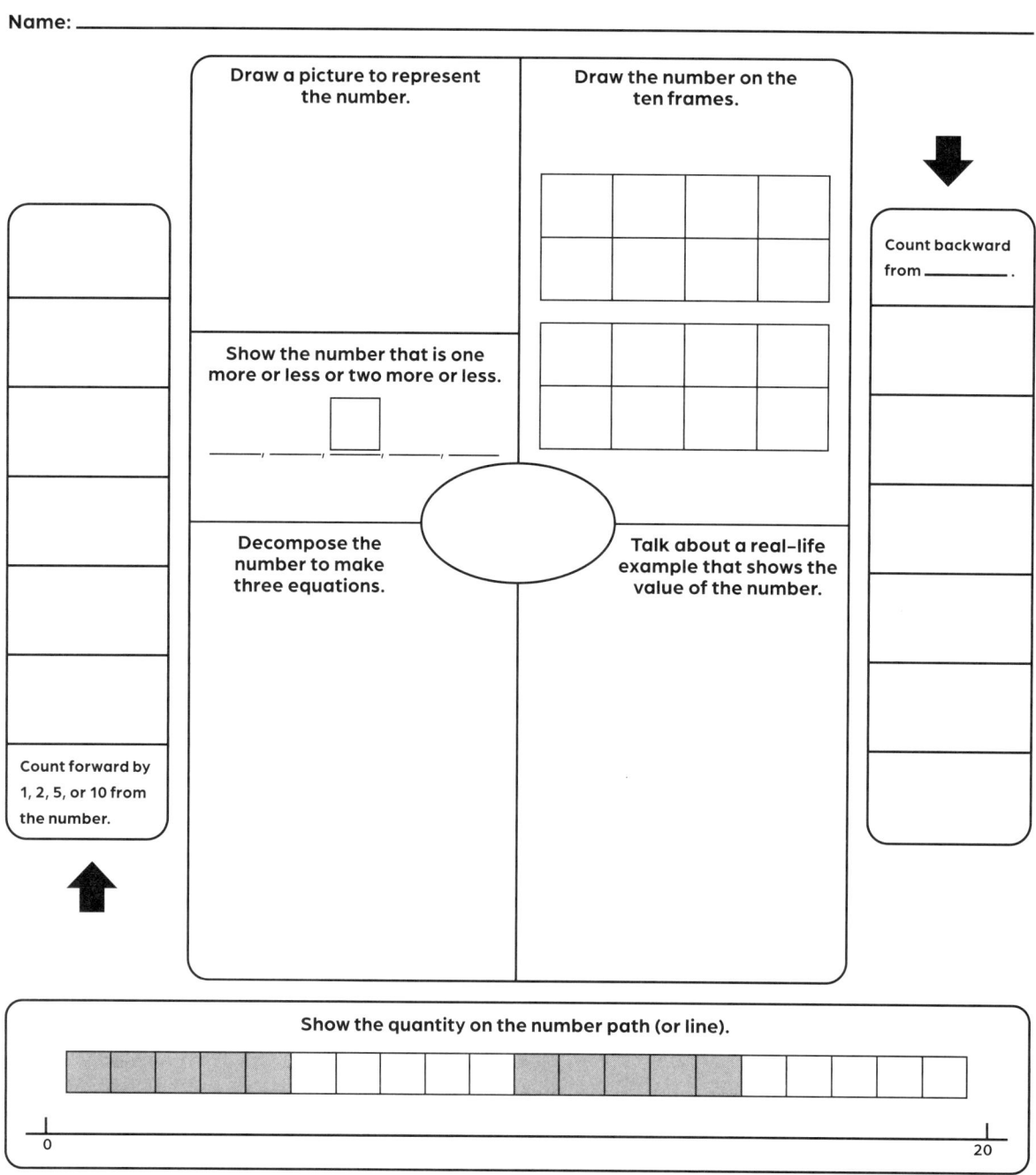

Source: Chilliwack School District (2016). Used with permission.

**FIGURE 4.4:** Grade 1 number sense SNAP.

Visit *go.SolutionTree.com/mathematics* for a free reproducible version of this figure.

## DRAW A PICTURE TO REPRESENT THE NUMBER

Drawing a picture to represent a number builds on many of the ideas teachers introduce to students in kindergarten and helps students conceptualize the value of numbers. Use tallying (for example, 5, 10, up to 20) as a doorway to understanding (see figure 4.5). Students then move on to drawing pictures of items (such as flowers or suns) and completing the hands-on activities (such as making balls or snakes out of clay). These are excellent ways for students to represent different numbers. When students work on the SNAP, using the tallying strategy can help because students sometimes reverse numbers in the tens. Teachers can make two or more groups, having students work in circle groups of ten.

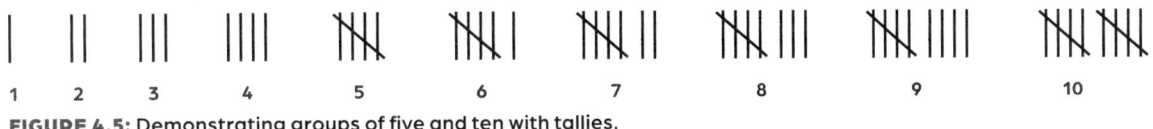

**FIGURE 4.5:** Demonstrating groups of five and ten with tallies.

## BUILD THE IDEAS OF FIVE AND TEN AND ONE MORE OR LESS

Teachers continue to teach and reinforce the concepts of five and ten and one more or less in grade 1. Building on their prior knowledge (what students learned in kindergarten), have students build sets of five, then show one more, and progress from there (see figure 4.5). Hands-on activities such as *loose parts* (objects students can count, sort, organize, and more) and using manipulatives to build sets are great ways to introduce this concept. Tallying is another effective way to demonstrate this concept. Whiteboards and document cameras are helpful for whole-class instruction, with students each using their whiteboard and then flipping it up to reveal their answers. Students need a great deal of practice to sequence numbers correctly, so teachers tell them, "You can do it if you can count it." The National Research Council (2001) explains when students progress from real-world scenarios to pictorial to verbal, and finally, to symbolic representation, they are developing conceptual knowledge of mathematical ideas. Therefore, as an ideal for student learning, move from concrete to representation to abstract. For example, in the concrete stage, teachers could ask students to locate five buttons from a box of objects. As a representation, teachers could ask students to draw five buttons. Moving from here to the abstract concept, teachers could ask students to perform a skill using the numeral 5.

## DECOMPOSE THE NUMBER TO MAKE EQUATIONS

Along with decomposing bags, another good strategy for learning this concept is to have students stand in groups of five and then have them split into subgroups (such as one and four or two and three). Students enjoy the movement aspect of this, and this activity also stresses that students can find or see mathematics taking place everywhere. *Rekenreks* (or counting frames) are another helpful way for students to decompose numbers and promote other pathways of understanding concerning the critical concept of part-part-whole (see figure 4.6, page 68).

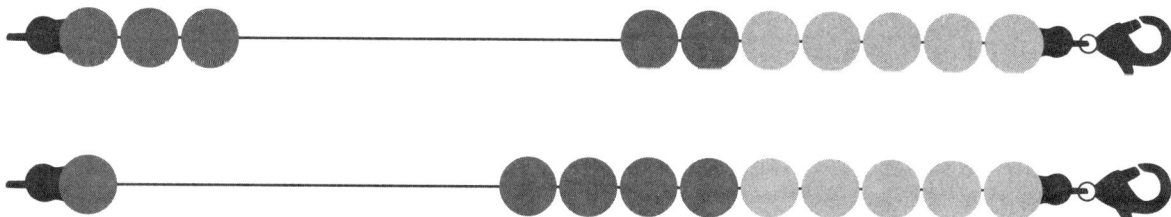

**FIGURE 4.6:** Using rekenreks to decompose numbers.

### DRAW THE NUMBER ON THE TEN FRAMES

Teachers introduce ten frames in kindergarten and their use continues to be effective in grade 1 and beyond (see figure 4.7). For example, a more advanced method of using ten frames could involve having students visualize the number 8, then have students close their eyes, and ask, "How many frames are filled in on the top, and how many are filled in on the bottom?" Students enjoy playing with number frames and are eager to see and build a number. Once students finish building a number, it is important to have them describe the process to a partner or their group.

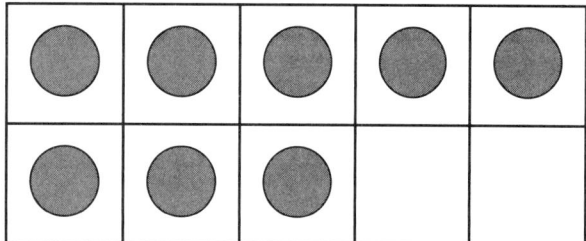

**FIGURE 4.7:** Using a ten frame to represent the number 8.

### TALK ABOUT A REAL-LIFE EXAMPLE THAT SHOWS THE NUMBER

Many teachers have heard students say, "How does this mathematics relate to real life?" Students at an early age need to see that mathematics is everywhere; they can find numbers in real life. To introduce this idea, have students begin with something familiar, such as what they eat. Pose questions such as, "What food can you eat five of and not get a tummy ache? Slices of pizza? Or strawberries?" Another question could be, "What can you hold in your hand?" Have students observe different objects around the classroom and ask them to picture the school spaces, such as three doors at the front of the building, ten soccer balls in a bag in the gym, and so on. These are just some of the ways teachers can reinforce the concept of numbers in real life.

### COUNT FORWARD AND BACKWARD

Building on what students already know, use the *morning meeting* strategy to work on counting forward and backward. Morning meetings are an effective way for teachers to promote social and emotional learning by connecting with students at the beginning of the day. Typically, morning meetings take place in a circle, where students each share how they are feeling or their goals for the day. Placing a number line from 1–20 above a calendar works well as a strategy for having students count forward and backward. Talking about "the number of the day" at the morning meeting is also effective. While taking attendance, teachers can review how many students are in the class, and use games such as *Around the World*, where students count off while in a circle, to effectively reinforce this concept. Providing *counting collections* (or different containers filled with different objects) is another great way for students to work on counting. Teachers can ask students to estimate how many items are in the container, and then have them count the items both forward and backward, which is another way of practicing counting (see figure 4.8).

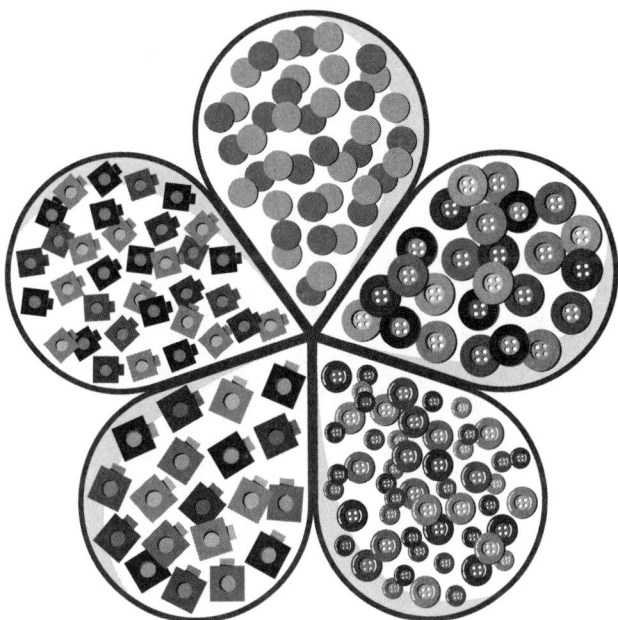

**FIGURE 4.8:** Counting collections.

## SHOW THE QUANTITY ON THE NUMBER LINE

Notice how the number line is divided into groups of five to promote benchmarks of five and ten in figure 4.9. Again, students enjoy it when teachers provide non-examples; for example, instruction could begin with the number 3. The teacher could color four boxes, and when the students notice the error, they can try to justify their thinking. Students must understand the spatial correspondence between numbers relating to their size. If students can print numbers from 0–20, count them orally, and represent them, can they show the quantity?

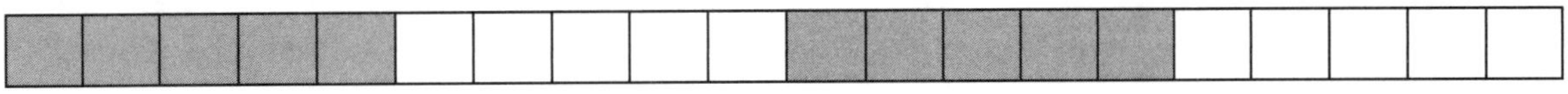

**FIGURE 4.9:** Number line in grade 1.

## Chapter Summary

We cannot stress enough how important it is to instill the concept of number sense in the first two years of school. A pedagogy immersed in curiosity, where teachers scaffold and interweave instruction, is fundamental for teachers to consider. Having a strong foundation in number sense not only allows students to explore the curriculum with confidence in the early grades but also prepares them for new challenges as they progress through the higher grades. Being able to subitize, understand part-part-whole, and explain the value of a number, together with the ability to reflect on learning, will go a long way in developing independent learners. In chapter 5, we talk about how to use the SNAP rubric with individual students and an entire class.

# CHAPTER 5

# Exploring Rubrics, Assessment, and Competency-Based Learning

In this chapter, we discuss the details of assessment through the lens of a teacher implementing the SNAP as a formative, diagnostic, or summative assessment. We describe how teachers can best interact with the SNAP to determine students' achievement levels and how to interpret what to do next.

Because the SNAP is a competency-based assessment, there are no distinct right or wrong answers. Teachers must value and consider student responses on a spectrum of possibilities. For example, there are infinite ways students might deconstruct a number and put it back together. These possibilities range from mundane and basic to profound and insightful. Teachers then determine the value of each option depending on students' ability to represent their thinking clearly and support it with evidence. The SNAP shines here because it is a window into the working minds of students. It is also a mirror—students can reflect on their learning, and teachers can plan for enrichment. This chapter introduces the SNAP rubric and explains how to use it with individual students and the whole class.

## Understanding the SNAP Rubric

We created the SNAP rubric to assist teachers with making sense of the spectrum of thinking approaches and strategies students will use. Much like a legend on a road map, this rubric fully aligns with the SNAP competencies, and we are confident these competencies will align with any school system curriculum.

Column three (*Applying*) of the SNAP rubric provides exemplar language of proficiency (see figure 5.1). The other achievement levels' columns (*Emerging*, *Developing*, and *Extending*) are left blank, allowing teachers to use their professional judgment to fill in evidence. Because the SNAP is a holistic assessment, teachers must consider the entire student work when looking for student understanding. For instance, if a student has responded insightfully on a part of the page that the teacher hadn't anticipated, or if the student connected the real-life example to a novel the student was reading, the teacher should acknowledge the student's thinking and give credit.

## SNAP Number Sense Rubric

| Competency | 1—Emerging<br>Student understanding and application of learning standards are not evident. | 2—Developing<br>The student demonstrates some understanding and application of number sense. | 3—Applying<br>The student demonstrates proficient understanding and application of number sense. | 4—Extending<br>The student demonstrates insightful understanding and application. |
|---|---|---|---|---|
| Conceptual understanding<br>*Picture box* | | | • Picture clearly communicates and represents the value of the number<br>• Accurate | |
| Describe picture | | | • Accurately describes and elaborates on pictorial representation<br>• Clearly communicates | |
| Procedural fluency<br>*Expanded form* | | | • Accurately demonstrates the value of each digit | |
| Counting forward and backward | | | • Complete and accurate | |
| Strategic competence<br>*Real-life connection* | | | • Provides connection to a real-life example<br>• Demonstrates understanding of the number value | |
| Three equations | | | • Accurately uses grade-appropriate operations in all three equations | |
| Adaptive reasoning<br>*Number line* | | | • Correctly estimates placement of number on provided number line with benchmarks (justification) | |
| Productive disposition<br>*Reflection* | | | • Some insight on mathematical thinking evident<br>• Evidence of self-efficacy in response | |
| **Cut Scores:** | 0–7: Emerging | 8–12: Developing | 13–16: Applying | 17–20: Extending |

*Source: Chilliwack School District (2016). Used with permission.*

**FIGURE 5.1:** Number sense SNAP rubric.

*Visit **go.SolutionTree.com/mathematics** for a free reproducible version of this figure.*

In the following sections, we provide two examples of a class profile to showcase what a teacher might experience and how the teacher might respond by using a classwide SNAP. We use the SNAP number sense class profile template (see figure 5.2) for both examples. For the first example, we use a mixed-number task and for the second, a three-digit rational number task in the section "Class Profile Example of a Mixed-Number Task."

## SNAP Number Sense Class Profile

Teacher: _____ Year: _____ Doorway question: _____

| Conceptual understanding | Procedural fluency | Strategic competence | Adaptive reasoning | Productive disposition |
|---|---|---|---|---|
| *Picture box and description picture* | *Expanded form and counting forward and backward* | *Real-life connection and three equations* | *Number line* | *Reflection* |
| Not evident   Proficient | Not evident   Proficient | Not evident   Proficient | Not evident   Proficient | Not evident   Proficient |

| Whole-class instruction | Small-group instruction | Individual instruction |
|---|---|---|
| | | |

*Source: Chilliwack School District (2016). Used with permission.*

**FIGURE 5.2:** The SNAP number sense class profile template.

*Visit **go.SolutionTree.com/mathematics** for a free reproducible version of this figure.*

## Class Profile Example of a Mixed-Number Task

A curricular outcome for upper-intermediate levels (grades 4–7, depending on jurisdiction) is students must understand fractions to show proficiency. For example, a teacher asked students to show what they knew about the mixed number 5 ¼ and students completed the SNAP task in roughly fifteen minutes. Here, we dig into three students' responses and analyze where they demonstrate proficiency and where they require support.

**STUDENT A**

Figure 5.3 shows student A's responses on the SNAP.

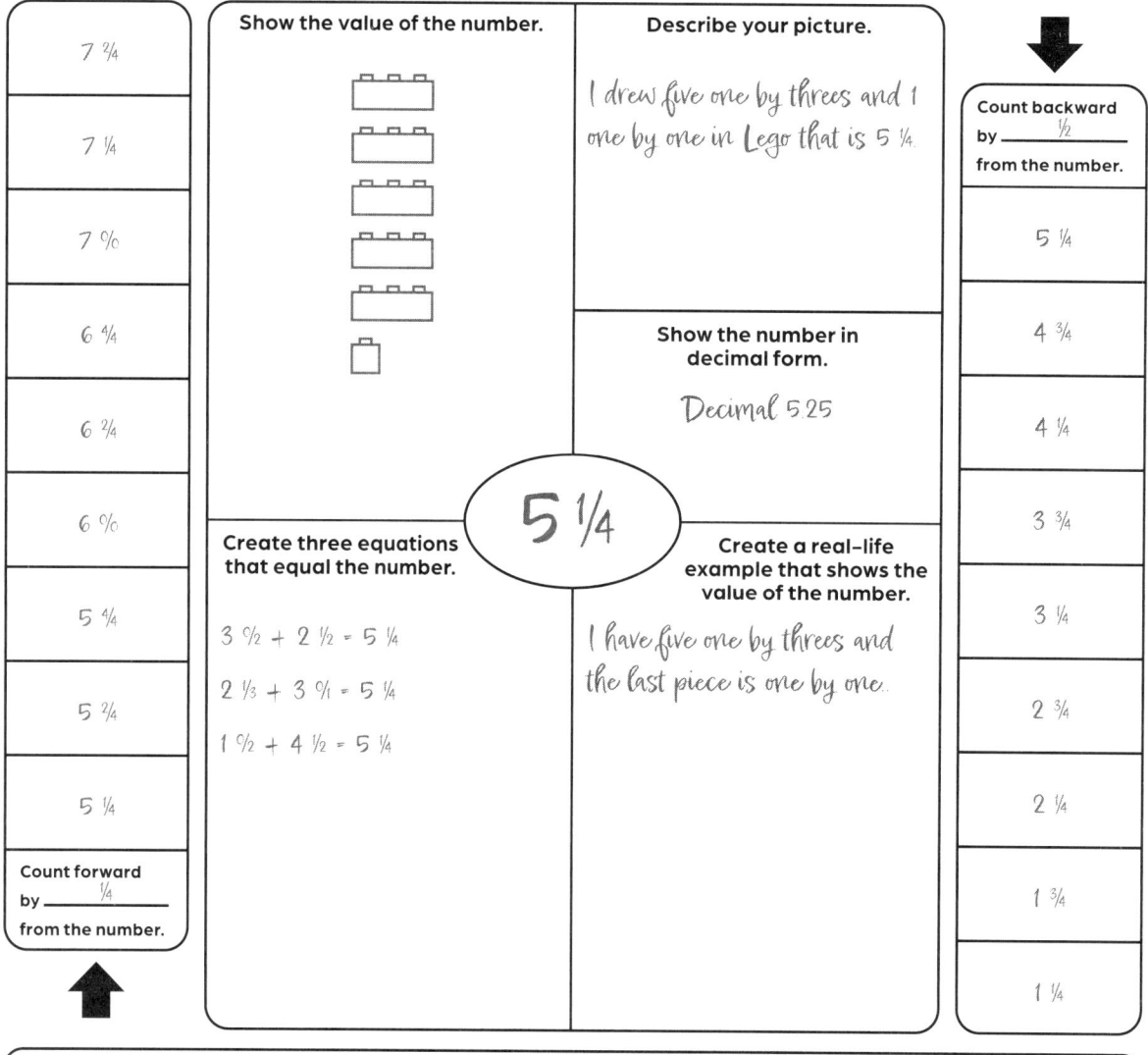

*Source: Chilliwack School District (2016). Used with permission.*
**FIGURE 5.3:** Student A's responses on the SNAP.

**Draw to represent:** Student A drew five wholes with three parts in each, plus a final one-third part. It is clear the student has some understanding of fractions, but has some confusion about a common denominator of fourths. (*Emerging* or *Developing*)

**Write to describe:** The student provided some context that further supports the student's confusion in the drawing. (*Developing*)

**Write in expanded (converted to decimal) form:** The student converted this correctly. (*Proficient*)

**Create three equations:** The student could not complete any equations correctly. The student valiantly made three attempts, but demonstrated an ineffective strategy with how to add fractions. (*Emerging*)

**Real-life example:** The student used Legos to describe understanding of 5 ¼. Unfortunately, the student mixed up incorrect pieces, and the example does not match the student's drawing. (*Emerging* or *Developing*)

**Counting forward by ¼:** The student shows lots of confusion here. Portions of the student's work are correct, but other portions are not. (*Emerging* or *Developing*)

**Counting backward by ½:** The student did this correctly. (*Proficient*)

**Number line:** The student randomly placed the number on the number line the teacher provided without further context. (*Emerging*)

**Reflection:** The student provided a surface response: "It was fairly easy." The student does not demonstrate self-reflection or the realization that the student's understanding is incomplete. (*Emerging*)

**Overall:** Student A needs significant support to show proficiency for this outcome. Small-group instruction will benefit this student in the following areas.

- Representation
- Equation work
- Real-life examples
- Counting forward and backward
- Number line work
- Metacognitive strategies

**Recommendations:** Student A needs some differentiated assessment to delve into areas of concern. We suggest trying the fraction ⅖ as the doorway question on a second SNAP. This will help show whether the student's confusion stems from the idea of a fraction or the complexity of the mixed number. Either way, the direction forward is clear, and we are confident the teacher can help improve student A's mathematics skills and understanding.

**Student A rubric example:** Figure 5.4 (page 76) is an example of a completed rubric for student A based on the SNAP. Note student A has areas of strength and stretches (see highlighted text in figure 5.4). The highlighted text in the Emerging and

## SNAP Number Sense Rubric

| Competency | 1—Emerging<br>*Student understanding and application of learning standards are not evident.* | 2—Developing<br>*The student demonstrates some understanding and application of number sense.* | 3—Applying<br>*The student demonstrates proficient understanding and application of number sense.* | 4—Extending<br>*The student demonstrates insightful understanding and application.* |
|---|---|---|---|---|
| Conceptual understanding<br>*Picture box* | | The picture clearly communicated, but inaccurately | • Picture clearly communicates and represents the value of the number<br>• Accurate | |
| Describe picture | | | • Accurately describes and elaborates on pictorial representation<br>• Clearly communicates | |
| Procedural fluency<br>*Expanded form* | | | • Accurately demonstrates the value of each digit | |
| Counting forward and backward | | Can count forward, but not backward | • Complete and accurate | |
| Strategic competence<br>*Real-life connection* | | Some misunderstandings evident | • Provides connection to a real-life example<br>• Demonstrates understanding of the number value | |
| Three equations | | Equations incorrect | • Accurately uses grade-appropriate operations in all three equations | |
| Adaptive reasoning<br>*Number line* | Made errors on number line; no benchmarks provided | | • Correctly estimates placement of number on provided number line with benchmarks (justification) | |
| Productive disposition<br>*Reflection* | Only provided a surface response | | • Some insight on mathematical thinking evident<br>• Evidence of self-efficacy in response | |
| **Cut Scores:** | 0–7: Emerging | 8–12: Developing | 13–16: Applying | 17–20: Extending |

*Source: Chilliwack School District (2016). Used with permission.*
**FIGURE 5.4:** Number sense rubric for student A.

Developing columns are written teacher comments, whereas the text in the Applying column is standard language from the rubric template (see figure 5.1, page 72). Once the teacher completes a rubric for each student, the teacher can then use that information to complete the class profile. The class profile provides a visual overview of the strengths and stretches of all students in the classroom, and features where students require whole-class, small-group, and individual instruction (see figure 5.2, page 73).

**STUDENT B**

Figure 5.5 shows student B's responses on the SNAP.

# SNAP
## Number Sense
**0–10,000**

**Name:** Student B

**Date:**

**Show the value of the number.**

**Describe your picture.**

= 1

= 1/4

**Show the number in decimal form.**

Decimal 5.25

**5 1/4**

**Create three equations that equal the number.**

5 + 1/4 = 5 1/4

6 − 3/4 = 5 1/4

4 + 5/4 = 5 1/4

**Create a real-life example that shows the value of the number.**

I bought a pizza they cut it into 6 slices and I ate 5 1/4 of the pizza.

**Count forward by** 1/4 **from the number.**

| |
|---|
| 7 1/4 |
| 7 |
| 6 3/4 |
| 6 2/4 |
| 6 1/4 |
| 6 |
| 5 3/4 |
| 5 2/4 |
| 5 1/4 |

**Count backward by** 1/2 **from the number.**

| |
|---|
| 5 1/4 |
| 5 3/4 |
| 6 1/4 |
| 6 3/4 |
| 7 1/4 |
| 7 3/4 |
| 8 1/4 |
| 8 3/4 |
| 9 1/4 |

**Show where the number falls on the number line.**

0 1 2 3 4 5 5 1/4 6 7 8 9 10

**Reflect**

It was hard.

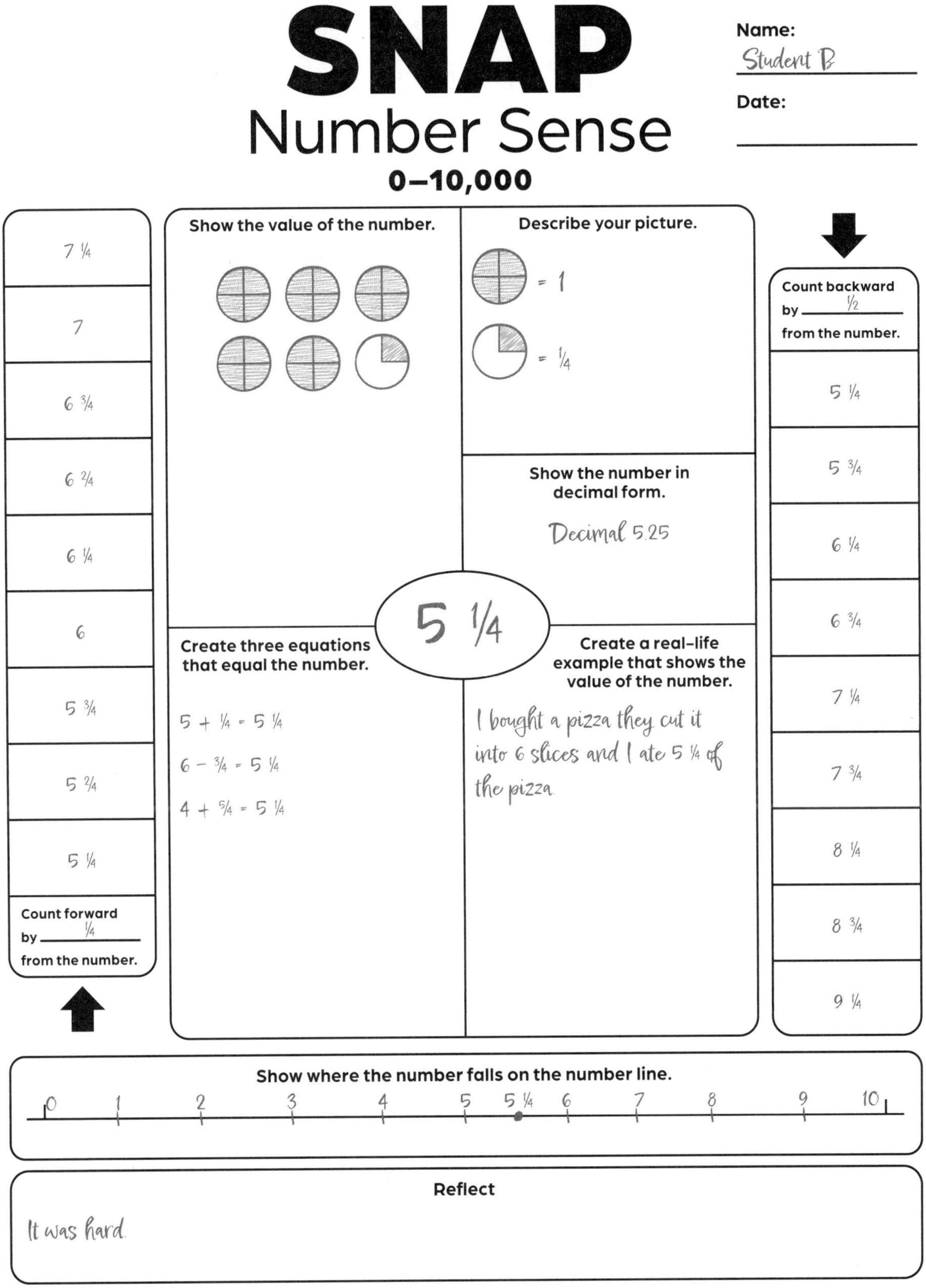

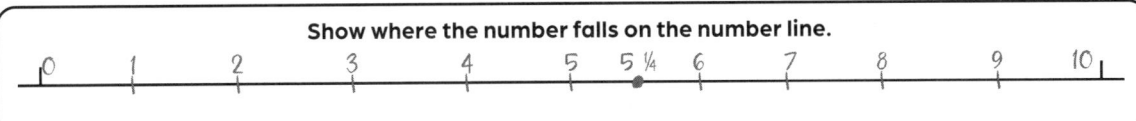

*Source: Chilliwack School District (2016). Used with permission.*

**FIGURE 5.5:** Student B's responses on the SNAP.

**Draw to represent:** Student B drew five wholes with four parts in each, plus a final ¼ part. No concerns. (*Proficient*)

**Write to describe:** The student provided some context via a legend that provides context to the drawing. (*Proficient*)

**Write in expanded (converted to decimal) form:** The student converted this correctly. (*Proficient*)

**Create three equations:** The student completed the three equations correctly and appropriately. (*Proficient*)

**Real-life example:** It's interesting that the student showed good understanding up to this point in the assessment. However, in the real-life example, the student showed some misunderstandings. (*Developing*)

**Counting forward by ¼:** No issues. (*Proficient*)

**Counting backward by ½:** The student cannot count backward, only forward. Did the student misunderstand the task? It would be worth doing a verbal check-in. However, based on what the student wrote, the student needs more work on this skill. (*Developing*)

**Number line:** The student drew a number line with an appropriate origin and end point. However, the student placed the ¼ point at the halfway point. This shows some confusion and is an area to improve. (*Developing*)

**Reflection:** The student provided a surface response, "It was hard." There is not much self-reflection evident in the response. (*Emerging*)

**Overall:** Student B has some strengths in this number sense task. However, there are also areas of concern where the student needs support. Small-group instruction will benefit this student in the following areas.

- Real-life examples
- Counting backward
- Number line work
- Metacognitive strategies

**Recommendations:** Student B needs differentiated assessment to support the student's learning. The level of concern for this student is not high since the student does have some strong skills to lean on. Based on the SNAP, it is possible the student may only need short interventions to get on the right track. For instance, the student's number line work and counting backward are close to correct. A one-to-one conversation may clear up those areas. More intervention will be necessary in the areas of real-life examples and metacognitive strategies.

Figure 5.6 is the completed rubric for student B based on the SNAP.

## SNAP Number Sense Rubric

| Competency | 1—Emerging<br>*Student understanding and application of learning standards are not evident.* | 2—Developing<br>*The student demonstrates some understanding and application of number sense.* | 3—Applying<br>*The student demonstrates proficient understanding and application of number sense.* | 4—Extending<br>*The student demonstrates insightful understanding and application.* |
|---|---|---|---|---|
| Conceptual understanding<br>*Picture box* | | | • Picture clearly communicates and represents the value of the number<br>• Accurate | |
| *Describe picture* | | | • Accurately describes and elaborates on pictorial representation<br>• Clearly communicates | |
| Procedural fluency<br>*Expanded form* | | | • Accurately demonstrates the value of each digit | |
| *Counting forward and backward* | | Can count forward, but not backward | • Complete and accurate | |
| Strategic competence<br>*Real-life connection* | | Some misunderstandings evident | • Provides connection to a real-life example<br>• Demonstrates understanding of the number value | |
| *Three equations* | | | • Accurately uses grade-appropriate operations in all three equations | |
| Adaptive reasoning<br>*Number line* | | Has appropriate origin and end point, but places 5 ¼ at the halfway mark. | • Correctly estimates placement of number on provided number line with benchmarks (justification) | |
| Productive disposition<br>*Reflection* | Only provided a surface response | | • Some insight on mathematical thinking evident<br>• Evidence of self-efficacy in response | |
| **Cut Scores:** | 0–7: Emerging | 8–12: Developing | 13–16: Applying | 17–20: Extending |

*Source: Chilliwack School District (2016). Used with permission.*

**FIGURE 5.6:** Number sense rubric for student B.

**STUDENT C**

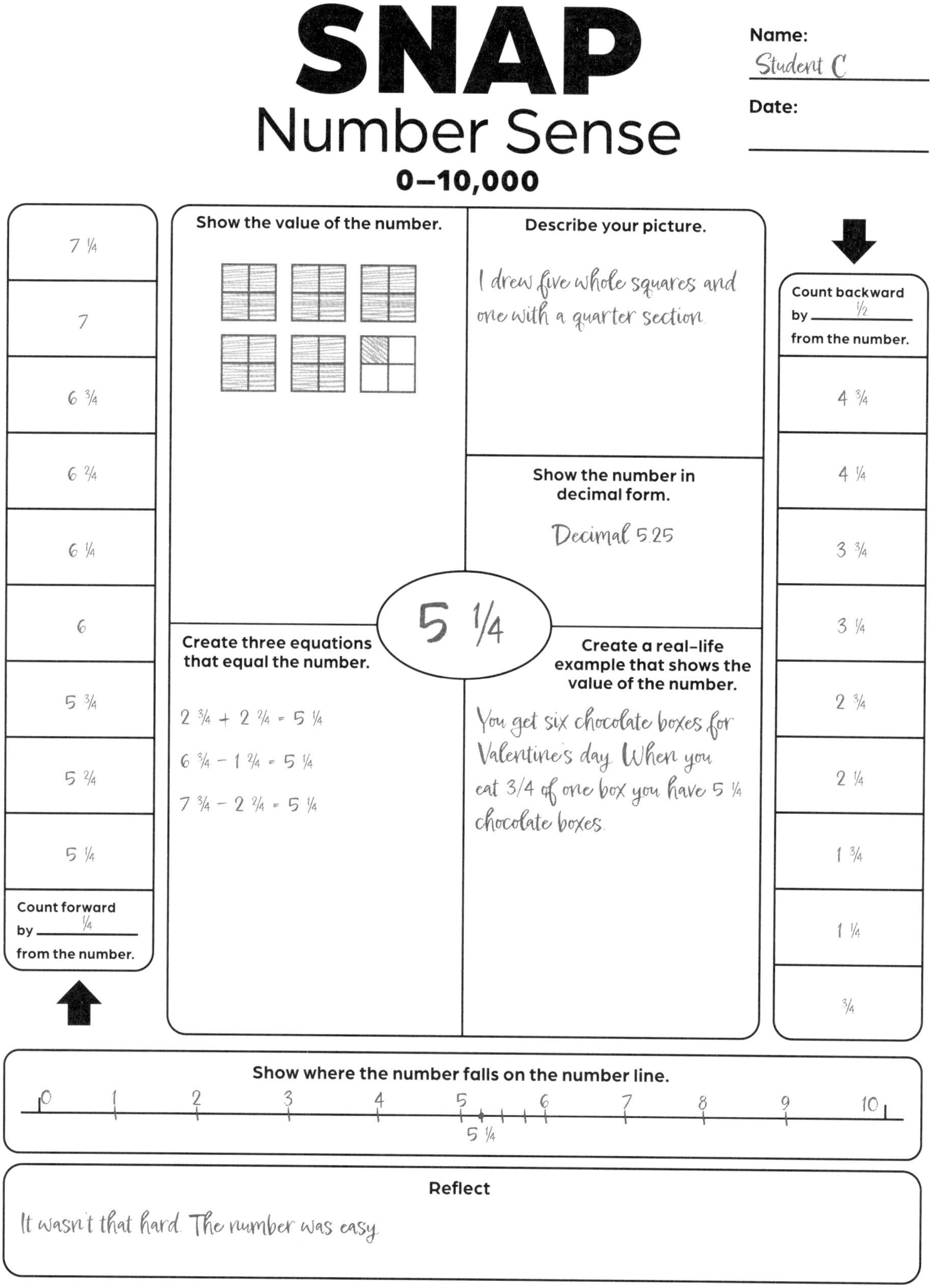

Figure 5.7 shows student C's responses on the SNAP.

SNAP

Number Sense

0–10,000

Name: Student C

Date:

Show the value of the number.

Describe your picture.

I drew five whole squares and one with a quarter section.

Show the number in decimal form.

Decimal 5.25

5 1/4

Create three equations that equal the number.

2 3/4 + 2 2/4 = 5 1/4

6 3/4 − 1 2/4 = 5 1/4

7 3/4 − 2 2/4 = 5 1/4

Create a real-life example that shows the value of the number.

You get six chocolate boxes for Valentine's day. When you eat 3/4 of one box you have 5 1/4 chocolate boxes.

Count forward by 1/4 from the number.

5 1/4, 5 2/4, 5 3/4, 6, 6 1/4, 6 2/4, 6 3/4, 7, 7 1/4

Count backward by 1/2 from the number.

4 3/4, 4 1/4, 3 3/4, 3 1/4, 2 3/4, 2 1/4, 1 3/4, 1 1/4, 3/4

Show where the number falls on the number line.

0 1 2 3 4 5 6 7 8 9 10

5 1/4

Reflect

It wasn't that hard. The number was easy.

*Source: Chilliwack School District (2016). Used with permission.*

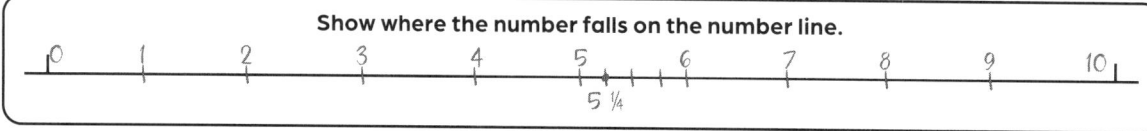

**FIGURE 5.7:** Student C's responses on the SNAP.

**Draw to represent:** Student C drew five wholes with four parts in each, plus a final ¼ part. No concerns. (*Proficient*)

**Write to describe:** The student explained the drawing and it makes sense. (*Proficient*)

**Write in expanded (converted to decimal) form:** The student converted this correctly. (*Proficient*)

**Create three equations:** The student completed the three equations correctly and appropriately. (*Proficient*)

**Real-life example:** The student used a plausible real-life example with chocolate boxes. (*Proficient*)

**Counting forward by ¼:** No issues. (*Proficient*)

**Counting backward by ½:** No issues. (*Proficient*)

**Number line:** The student drew a number line with an appropriate origin and end point and provided appropriate segmentation between 5 and 6 to show where 5 ¼ lies. (*Proficient*)

**Reflection:** The student responded, "It wasn't that hard. The number was easy." Based on the student's performance, this statement is true. However, the student's response lacks depth. (*Developing*)

**Overall:** Student C is proficient in this number sense task. There are no areas of concern other than the student needs to continue work on metacognitive strategies.

**Student C recommendations:** Student C has a good grasp of this number sense task. The teacher could push the student to dig deeper in the reflection and provide a more advanced doorway question (or the number in the middle) to challenge the student further and see if the student's knowledge is deeply entrenched and transferable.

Figure 5.8 (page 82) is an example of a completed rubric for student C based on the SNAP.

## CLASS PROFILE RECOMMENDATIONS FOR MIXED-NUMBER TASK

**Whole-class instruction:** Although we only go in depth on three students' assessments, some themes are beginning to emerge. These themes help teachers plan for instruction and ensure all students receive the instruction and support they need. See the sample class profile sheet in figure 5.9 (page 83). All three students (A, B, and C) are struggling with the reflection work. Their responses are thin and lack depth and understanding of their own perceptions of learning. Some prompts for students could include, What was easy? What was difficult? or What do I need to learn more about? This information helps the teacher plan for whole-class instruction in this area.

**Small-group instruction:** A guaranteed way to turn off students in a learning environment is to bore them with information and instruction they already know. Therefore, targeted, small-group instruction is helpful for the student-centered teacher. Based on the class profile (in figure 5.9), some groupings are Emerging. Students A and B show

| \multicolumn{5}{c}{**SNAP Number Sense Rubric**} |
| --- | --- | --- | --- | --- |
| **Competency** | **1—Emerging** *Student understanding and application of learning standards are not evident.* | **2—Developing** *The student demonstrates some understanding and application of number sense.* | **3—Applying** *The student demonstrates proficient understanding and application of number sense.* | **4—Extending** *The student demonstrates insightful understanding and application.* |
| Conceptual understanding *Picture box* | | | • Picture clearly communicates and represents the value of the number<br>• Accurate | |
| *Describe picture* | | | • Accurately describes and elaborates on pictorial representation<br>• Clearly communicates | |
| Procedural fluency *Expanded form* | | | • Accurately demonstrates the value of each digit | |
| *Counting forward and backward* | | | • Complete and accurate | |
| Strategic competence *Real-life connection* | | | • Provides connection to a real-life example<br>• Demonstrates understanding of the number value | |
| *Three equations* | | | • Accurately uses grade-appropriate operations in all three equations | |
| Adaptive reasoning *Number line* | | | • Correctly estimates placement of number on provided number line with benchmarks (justification) | |
| Productive disposition *Reflection* | | Needs more depth in the reflection | • Some insight on mathematical thinking evident<br>• Evidence of self-efficacy in response | |
| **Cut Scores:** | 0–7: Emerging | 8–12: Developing | 13–16: Applying | 17–20: Extending |

*Source: Chilliwack School District (2016). Used with permission.*
**FIGURE 5.8:** Number sense rubric for student C.

deficiencies in their understanding of real-life connections and number line work. The teacher could pair these two students together and, using the number line, the teacher could strengthen their skip counting skills, which is also a struggle for both students.

**Individual instruction:** Student A needs a lot of support to meet teacher expectations and requires a triple dose of interventions. In other words, in addition to the high-quality Tier 1 instruction this student (and the rest of the class) receives, student A needs quality Tier 2 *and* quality Tier 3 instruction. It is often a challenge for teachers

## SNAP Number Sense Class Profile

Teacher: __Example Teacher__  Year: __2023__  Doorway question: __5 ¼__

| Conceptual understanding<br>*Picture box and description picture* | | Procedural fluency<br>*Expanded form and counting forward and backward* | | Strategic competence<br>*Real-life connection and three equations* | | Adaptive reasoning<br>*Number line* | | Productive disposition<br>*Reflection* | |
|---|---|---|---|---|---|---|---|---|---|
| Not evident | Proficient | Not evident | Proficient | Not evident | Proficient | Not evident | Proficient | Not evident | Proficient |
| Student A | Student B | Student A | Student B | Student A | Student B | Student A | Student B | Student A | Student E |
| | Student C | Student C (backward counting) | Student D | Student C (real life) | Student C (three equations) | Student C | Student D | Student B | Student M |
| | Student D | | Student F | | | Student K | Student E | Student C | |
| | Student E | Student O | Student G | Student J | Student D | Student N | Student F | Student D | |
| | Student F | | Student H | Student L | Student E | Student O | Student G | Student F | |
| | Student G | | Student I | | Student F | | Student H | Student G | |
| | Student H | | Student J | | Student G | | Student I | Student H | |
| | Student I | | Student K | | Student H | | Student J | Student I | |
| | Student J | | Student L | | Student I | | Student L | Student J | |
| | Student K | | Student M | | Student K | | Student M | Student K | |
| | Student L | | Student N | | Student M | | Student P | Student L | |
| | Student M | | Student P | | Student N | | | Student N | |
| | Student N | | | | Student O | | | Student O | |
| | Student O | | | | Student P | | | Student P | |
| | Student P | | | | | | | | |

**Whole–class instruction**
Almost the entire class is struggling with deep reflections. The whole class needs metacognitive work. Students E and M require an extension activity.

**Small-group instruction**
Students A, C, O: Expanded form and skip counting

Students A, C, J, L: Real-life connections and equations

Students A, C, K, N, O: Number line

**Individual instruction**
Student A requires individual instruction with creating representations of the number.

*Source: Chilliwack School District (2016). Used with permission.*
**FIGURE 5.9:** SNAP number sense class profile for mixed-number task example.

---

to schedule interventions for one student, but the instruction is essential both if teachers are gathering evidence to make plans and to avoid having the student fall further behind. Student A needs help with picture representations, skip counting, and creating equations that equal the mixed number. With targeted instruction, the teachers can fill in the learning gaps.

## Class Profile Example of a Whole-Number Task

The teacher asked students to show what they knew about the number 579. Students completed this task in roughly fifteen minutes. This is a curricular outcome (place value, skip counting, comparing and ordering numbers up to 1,000) for primary levels (usually at the grade 3 level, depending on jurisdiction). We provide three student examples. Here, we dig into each student's response and analyze where the three students demonstrate proficiency, and where they require support.

## STUDENT 1

Figure 5.10 shows student 1's responses on the SNAP.

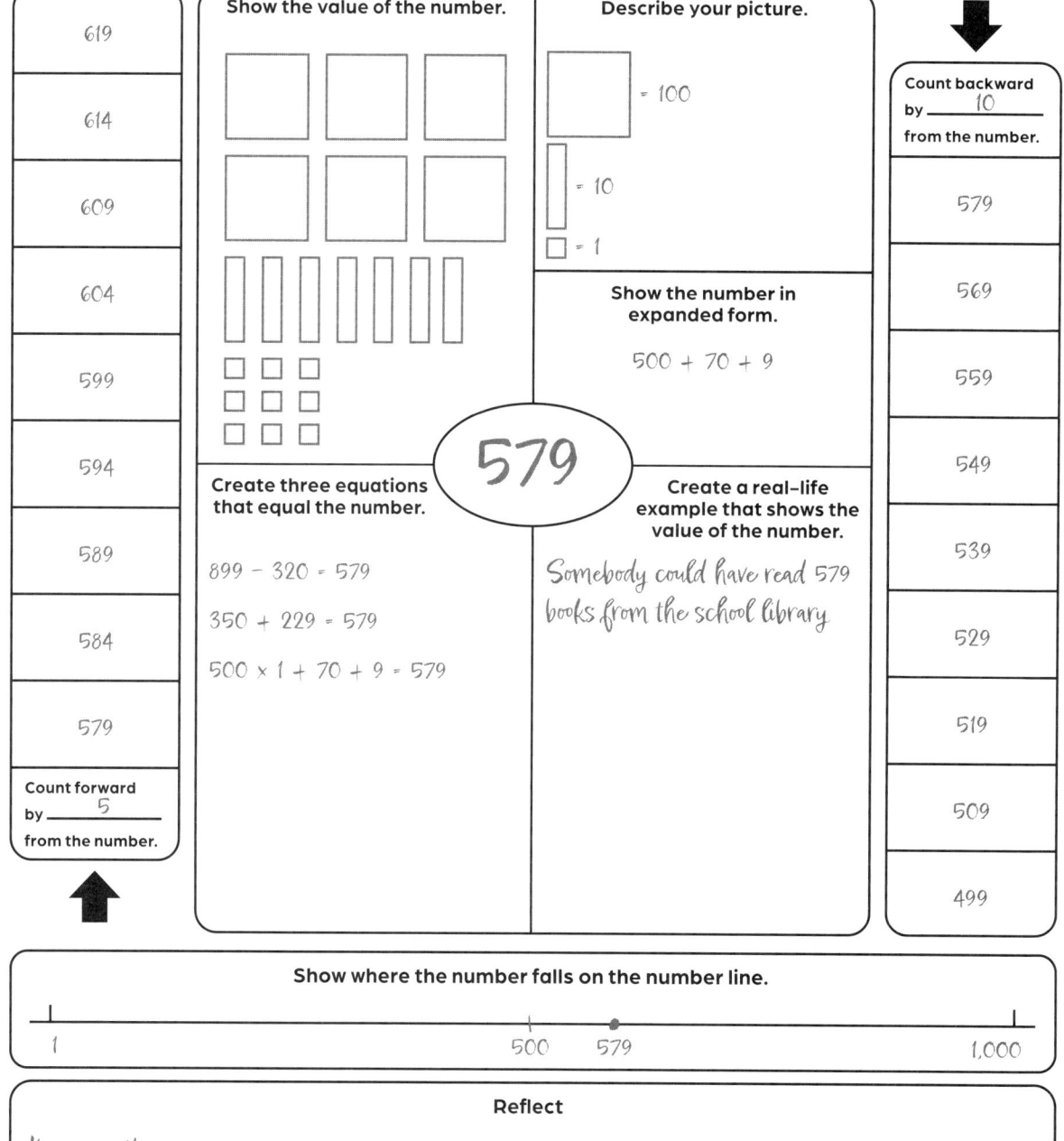

**FIGURE 5.10:** Student 1's responses on the SNAP.

**Draw to represent:** Student 1 drew five large squares, seven smaller rectangles, and nine small squares. When coupled with the student's description, this makes good sense. (*Proficient*)

**Write to describe:** The student provided context that further supports the understanding the student's drawing shows by providing a legend for each of the shapes. (*Proficient*)

**Write in expanded (converted to decimal) form:** The student converted this correctly. (*Proficient*)

**Create three equations:** The student completed this correctly. (*Proficient*)

**Real-life example:** The student wrote, "Somebody could have read 579 books from the school library." This is possible, but the student could have followed up with some extra context, which the student needs to validate this example. For instance, "A person *could* read 579 books from the school library if the person read one book a week. It would take the person almost 12 years!" (*Developing*, but could move to *Proficient* after a conversation)

**Counting forward by 5:** Correct. (*Proficient*)

**Counting backward by 10:** Correct. (*Proficient*)

**Number line:** The student drew an origin, end point, and midpoint on the number line. It appears that the student estimated where 579 would be located; it is a good estimate. No concerns. (*Proficient*)

**Reflection:** The student provided a surface response, "It was pretty easy." There is not much self-reflection, but based on the student's performance, it might have been easy. Still, teachers are expecting a little more context to achieve a proficient score. More context might include a statement as to *why* this was an easy task: "I used to have trouble with the number line, but since I learned how to find the midpoint of a number line, it is easy to place any number." (*Developing*)

**Overall:** Student 1 does not need significant support to show proficiency on this outcome. Small-group instruction will benefit this student in the area of metacognitive strategies.

**Student 1 recommendations:** Student 1 needs very little differentiated instruction to become fully proficient on this outcome. We suggest further instruction in metacognitive strategies, either with the whole class or a small group. Additionally, we recommend a conversation check-in regarding student 1's understanding of 579 in the real world to uncover more context and help to determine if further intervention is needed.

Figure 5.11 (page 86) is an example of a completed rubric for student 1 based on the SNAP.

| SNAP Number Sense Rubric | | | | |
|---|---|---|---|---|
| Competency | 1—Emerging<br>*Student understanding and application of learning standards are not evident.* | 2—Developing<br>*The student demonstrates some understanding and application of number sense.* | 3—Applying<br>*The student demonstrates proficient understanding and application of number sense.* | 4—Extending<br>*The student demonstrates insightful understanding and application.* |
| Conceptual understanding<br>*Picture box* | | | • Picture clearly communicates and represents the value of the number<br>• Accurate | |
| *Describe picture* | | | • Accurately describes and elaborates on pictorial representation<br>• Clearly communicates | |
| Procedural fluency<br>*Expanded form* | | | • Accurately demonstrates the value of each digit | |
| *Counting forward and backward* | | | • Complete and accurate | |
| Strategic competence<br>*Real-life connection* | | *This is a possible example, but it needs more context to justify the student's thinking. Reading 579 library books is a lot!* | • Provides connection to a real-life example<br>• Demonstrates understanding of the number value | |
| *Three equations* | | | • Accurately uses grade-appropriate operations in all three equations | |
| Adaptive reasoning<br>*Number line* | | | • Correctly estimates placement of number on provided number line with benchmarks (justification) | |
| Productive disposition<br>*Reflection* | | *The student could have provided more depth in the reflection.* | • Some insight on mathematical thinking evident<br>• Evidence of self-efficacy in response | |
| **Cut Scores:** | 0–7: Emerging | 8–12: Developing | 13–16: Applying | 17–20: Extending |

**FIGURE 5.11:** Number sense rubric for student 1.

**STUDENT 2**

Figure 5.12 shows student 2's responses on the SNAP.

# SNAP

## Number Sense

**0–10,000**

Name: Student 2

Date:

Show the value of the number.

100 200 300 400 500 600 700

579

Describe your picture.

Skip counted on a number line.

Show the number in expanded form.

500 + 79

579

Create three equations that equal the number.

578 + 1 = 579

580 − 1 = 579

579 − 1 + 1 = 579

Create a real-life example that shows the value of the number.

Our house number is 579.

Count forward by 5 from the number.

620

615

610

605

600

599

594

589

584

Count backward by 10 from the number.

569

559

549

539

529

519

509

500

490

Show where the number falls on the number line.

100 200 300 400 500 600 700

Reflect

It was easy to do the equations and it was hard to think of a real-life example.

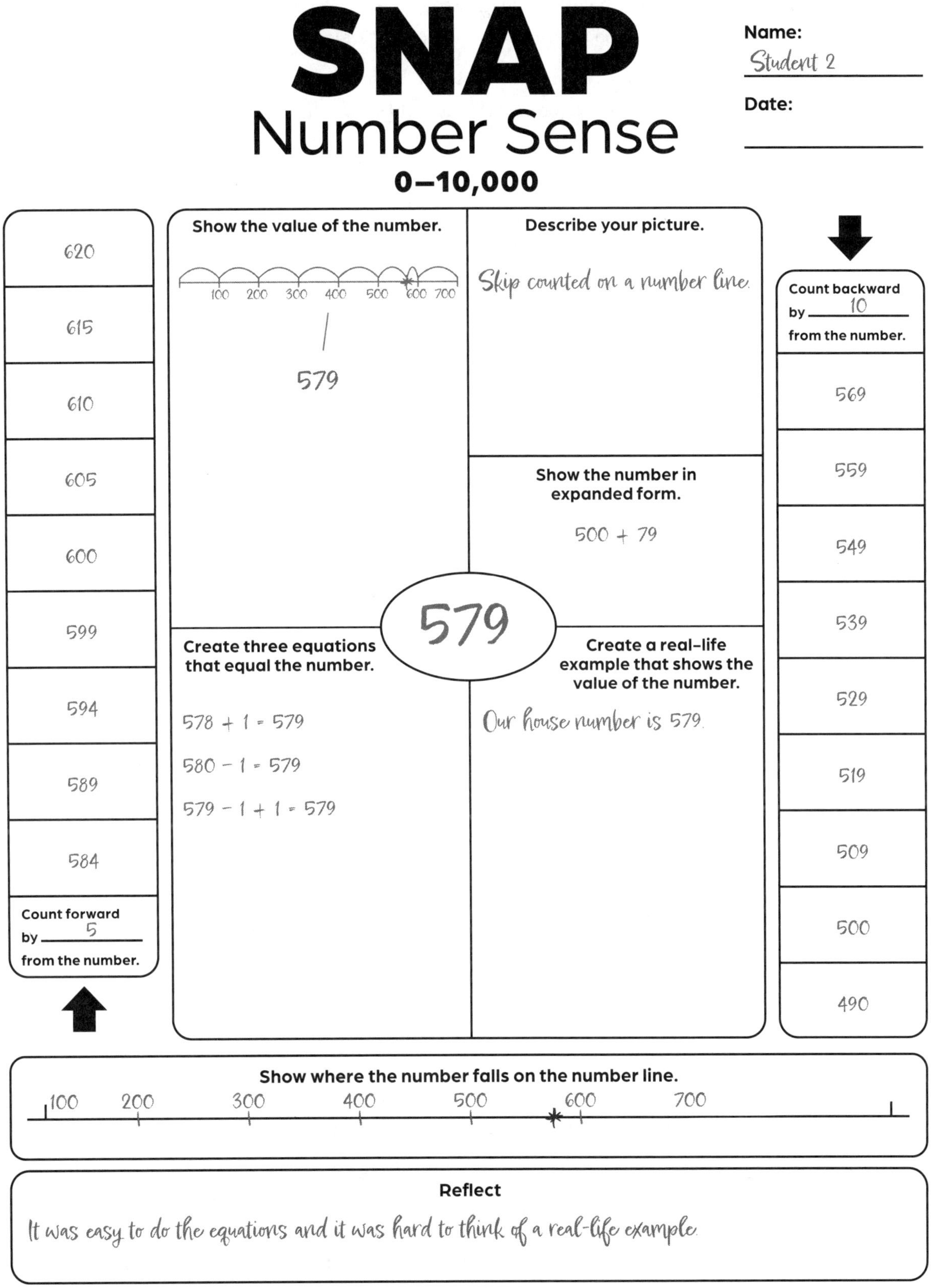

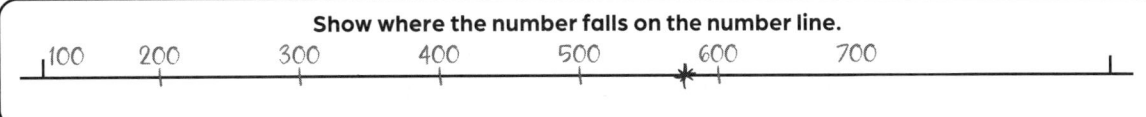

*Source: Chilliwack School District (2016). Used with permission.*

**FIGURE 5.12:** Student 2's responses on the SNAP.

**Draw to represent:** Student 2 drew a number line and skip counted correctly to 579. (*Proficient*)

**Write to describe:** The student wrote, "Skip counted on a number line." This is not much information, but minimally meets the standard. (*Proficient*)

**Write in expanded form:** The student did not fully complete this section. The student did not separate 70 + 9 and instead had 500 + 79. The student is on the right track, but not fully there. (*Developing*)

**Create three equations:** The student was able to complete this correctly, but in the simplest way. Adding one and taking away one is not grade-level appropriate. (*Developing*)

**Real-life example:** The student wrote, "Our house number is 579." This does not show an understanding of the value of the number. The student could provide more context, so the teacher can figure out what student 2 really knows. (*Developing*)

**Counting forward by 5:** Correct until the student crosses over from the 500s to 600s. (*Developing*)

**Counting backward by 10:** Correct until the student crosses over from the 500s to the 400s. (*Developing*)

**Number line:** The student drew 100 increments and labeled them on the number line. The student then placed 579 between 500 and 600. The student is missing a starting point and further delineations between 500 and 600 on the line, but the student gets the idea. (*Proficient*)

**Reflection:** The student provided some context about an easy and a difficult part of the assessment. At the grade 3 level, this does not demonstrate extended thinking, but it does meet the standard. (*Proficient*)

**Overall:** Student 2 shows strength and stretch areas on the SNAP. Small-group instruction will benefit this student in the following areas.

- Expanded form
- Create three equations
- Real-life example
- Counting forward and backward

**Student 2 recommendations:** Student 2 needs targeted and differentiated instruction to become fully proficient on this outcome. We suggest further instruction in the areas previously listed, particularly in the area of skip counting by 5 and 10 when crossing over to a new hundred. This is an odd area of confusion because the student was able to show moving from hundred to hundred on the number line. Certainly, this is an area for the teacher to follow up on, and we anticipate this confusion could be fixed quickly. By leaning on the strengths, student 2 should be able to shore up any misunderstandings on this outcome and move ahead without limitations.

Figure 5.13 is an example of a completed rubric for student 2 based on the SNAP.

*Exploring Rubrics, Assessment, and Competency-Based Learning*

## SNAP Number Sense Rubric

| Competency | 1—Emerging<br>Student understanding and application of learning standards are not evident. | 2—Developing<br>The student demonstrates some understanding and application of number sense. | 3—Applying<br>The student demonstrates proficient understanding and application of number sense. | 4—Extending<br>The student demonstrates insightful understanding and application. |
|---|---|---|---|---|
| Conceptual understanding<br>*Picture box* | | | • Picture clearly communicates and represents the value of the number<br>• Accurate | |
| *Describe picture* | | | • Accurately describes and elaborates on pictorial representation<br>• Clearly communicates | |
| Procedural fluency<br>*Expanded form* | | Did not separate 70 + 9; instead wrote 500 + 79 | • Accurately demonstrates the value of each digit | |
| Counting forward and backward | | Partially correct | • Complete and accurate | |
| Strategic competence<br>*Real-life connection* | | Example does not show an understanding of the value of the number | • Provides connection to a real-life example<br>• Demonstrates understanding of the number value | |
| Three equations | | Not a grade-appropriate example | • Accurately uses grade-appropriate operations in all three equations | |
| Adaptive reasoning<br>*Number line* | | | • Correctly estimates placement of number on provided number line with benchmarks (justification) | |
| Productive disposition<br>*Reflection* | | | • Some insight on mathematical thinking evident<br>• Evidence of self-efficacy in response | |
| **Cut Scores:** | 0–7: Emerging | 8–12: Developing | 13–16: Applying | 17–20: Extending |

**FIGURE 5.13:** Number sense rubric for student 2.

**STUDENT 3**

Figure 5.14 shows student 3's responses on the SNAP.

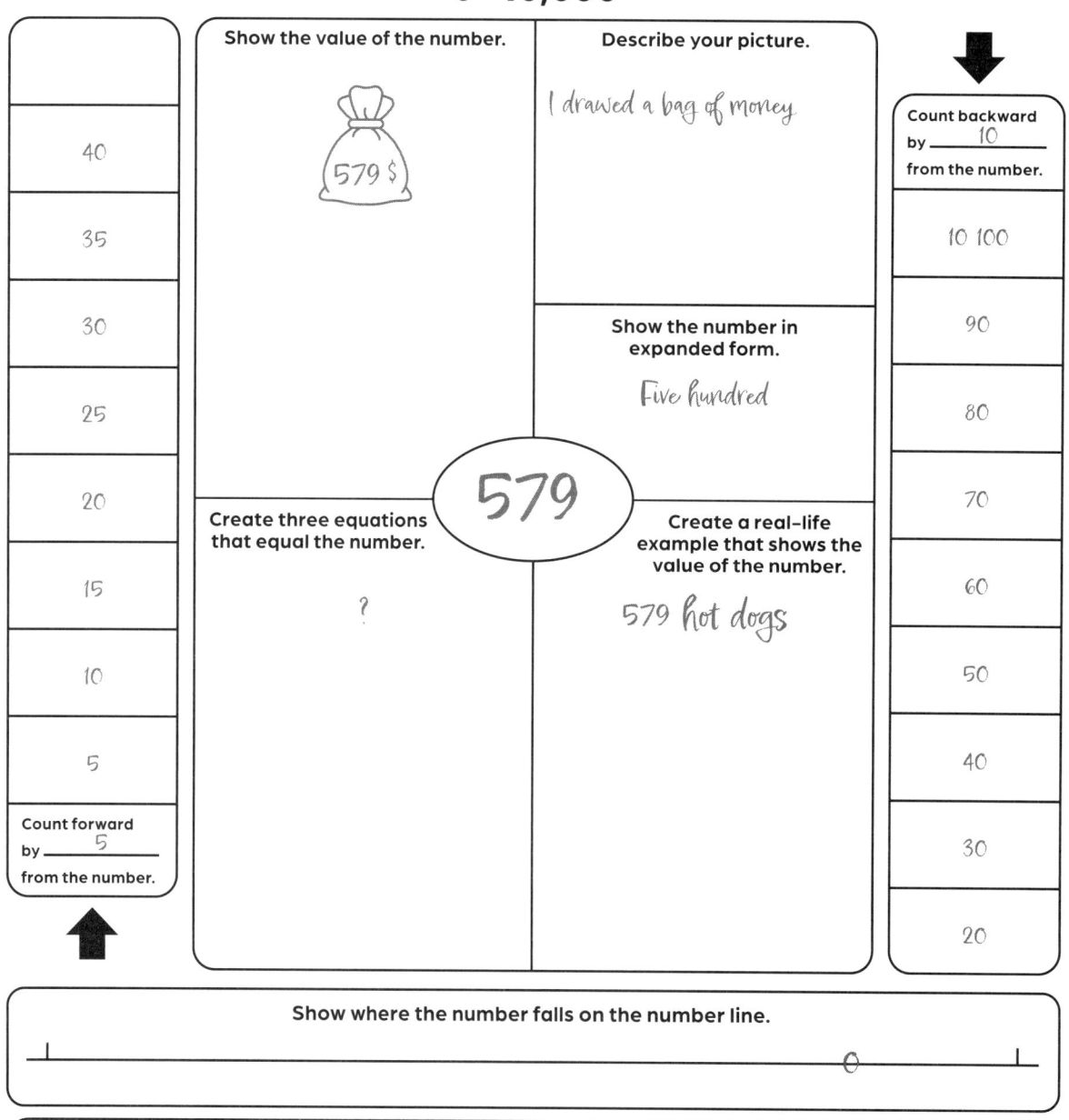

*Source: Chilliwack School District (2016). Used with permission.*
**FIGURE 5.14:** Student 3's responses on the SNAP.

**Draw to represent:** Student 3 drew a sack of money and labeled it "579 $." This drawing does not represent the value of the number and points to a lack of understanding. (*Emerging*)

**Write to describe:** The student wrote, "I drawed [*sic*] a bag of money." This does describe the picture but does not place the drawing in a mathematical context or meet the standard. (*Developing*)

**Write in expanded form:** The student did not fully complete this. The student only wrote "Five hundred"; this is not what the teacher wants the student to demonstrate here. (*Emerging*)

**Create three equations:** The student was unable to put anything here but a question mark, showing the student's confusion. (*Emerging*)

**Real-life example:** The student wrote "579 hot dogs." Whereas this could work as a real-life example with some additional context (for example, 579 hot dogs were sold at the school BBQ), the student's description needs more context for the teacher to figure out what student 3 really knows. (*Developing*)

**Counting forward by 5:** The student counted forward by 5, starting at 5. This was not the task, but it does show student 3 can count from 5 to 40 by fives. (*Developing*)

**Counting backward by 10:** The student started at 10, but then crossed out work (maybe because it would have required the student use negative integers). The student then restarted at 100 and counted backward correctly to 20. (*Developing*)

**Number line:** The student circled a portion of the number line without any referents or context. (*Emerging*)

**Reflection:** The student only wrote "I don't know." (*Emerging*)

**Overall:** Student 3 requires considerable support and intervention on this learning outcome, based on this SNAP. Small-group instruction will benefit this student in all areas of the assessment.

**Student 3 recommendations:** Student 3 needs targeted and differentiated instruction to become fully proficient on this outcome. We suggest the student complete a second SNAP with a smaller number to clarify what interventions the student needs. Perhaps try 79 as a starting number and see where the student continues to struggle, so the teacher can provide support in those areas first.

Figure 5.15 (page 92) is an example of a completed rubric for student 3 based on the SNAP.

| SNAP Number Sense Rubric |||||
|---|---|---|---|---|
| Competency | 1—Emerging<br>*Student understanding and application of learning standards are not evident.* | 2—Developing<br>*The student demonstrates some understanding and application of number sense.* | 3—Applying<br>*The student demonstrates proficient understanding and application of number sense.* | 4—Extending<br>*The student demonstrates insightful understanding and application.* |
| Conceptual understanding<br>*Picture box* | Drew a sack of money and labeled it "579 dollars" | | • Picture clearly communicates and represents the value of the number<br>• Accurate | |
| *Describe picture* | | Description does not place the drawing in a mathematical context. | • Accurately describes and elaborates on pictorial representation<br>• Clearly communicates | |
| Procedural fluency<br>*Expanded form* | Only wrote "500" | | • Accurately demonstrates the value of each digit | |
| *Counting forward and backward* | | Partially correct; uses the wrong value to skip count because needed to start with 579 | • Complete and accurate | |
| Strategic competence<br>*Real-life connection* | | Example does not show an understanding of the value of the number without more context | • Provides connection to a real-life example<br>• Demonstrates understanding of the number value | |
| *Three equations* | Not a grade-appropriate example | | • Accurately uses grade-appropriate operations in all three equations | |
| Adaptive reasoning<br>*Number line* | Circled a portion of the number line without any referents or context | | • Correctly estimates placement of number on provided number line with benchmarks (justification) | |
| Productive disposition<br>*Reflection* | Simply wrote "I don't know" | | • Some insight on mathematical thinking evident<br>• Evidence of self-efficacy in response | |
| **Cut Scores:** | 0–7: Emerging | 8–12: Developing | 13–16: Applying | 17–20: Extending |

**FIGURE 5.15:** Number sense rubric for student 3.

## CLASS PROFILE RECOMMENDATIONS FOR WHOLE-NUMBER TASK

**Whole-class instruction:** Based on this SNAP, two main areas of concern surface for this class.

1. **Real-life application:** Every student in the class is struggling with this concept. The teacher needs to fully attend to creating opportunities to bring the real world into the mathematics classroom. The teacher can accomplish this in various ways. We suggest using picture books, where the teacher stops at opportune moments in the story to create mathematical connections for students.

2. **Reflections:** All but two students show a need for extra teaching. We suggest the teacher model metacognition strategies, not just in mathematics class, but in all areas of the curriculum. Ideas for sentence starters include "I found _____ difficult today because _____," and "A helpful strategy I used to help me learn today was _____ because _____."

See figure 5.16 for the completed version of the SNAP number sense class profile for this task.

## SNAP Number Sense Class Profile

**Teacher:** Example Teacher   **Year:** 2023   **Doorway question:** 579

| Conceptual understanding | | Procedural fluency | | Strategic competence | | Adaptive reasoning | | Productive disposition | |
|---|---|---|---|---|---|---|---|---|---|
| *Picture box and description picture* | | *Expanded form and counting forward and backward* | | *Real-life connection and three equations* | | *Number line* | | *Reflection* | |
| Not evident | Proficient | Not evident | Proficient | Not evident | Proficient | Not evident | Proficient | Not evident | Proficient |
| Student 3 | Student 1 | Student 2 | Student 1 | Student 1 | | Student 3 | Student 1 | Student 1 | Student 2 |
| Student 5 | Student 2 | Student 3 | Student 4 | Student 2 | | Student 6 | Student 2 | Student 3 | Student 8 |
| Student 7 | Student 4 | Student 10 | Student 5 | Student 3 | | Student 7 | Student 4 | Student 4 | Student 9 |
| Student 10 | Student 6 | Student 13 | Student 6 | Student 4 | | Student 8 | Student 5 | Student 5 | |
| | Student 8 | Student 16 | Student 7 | Student 5 | | Student 10 | Student 9 | Student 6 | |
| | Student 9 | | Student 8 | Student 6 | | Student 13 | Student 10 | Student 7 | |
| | Student 11 | | Student 9 | Student 7 | | Student 16 | Student 11 | Student 10 | |
| | Student 12 | | Student 11 | Student 8 | | | Student 12 | Student 11 | |
| | Student 13 | | Student 12 | Student 9 | | | Student 14 | Student 12 | |
| | Student 14 | | Student 14 | Student 10 | | | Student 15 | Student 13 | |
| | Student 15 | | Student 15 | Student 11 | | | | Student 14 | |
| | Student 16 | | | Student 12 | | | | Student 15 | |
| | | | | Student 13 | | | | Student 16 | |
| | | | | Student 14 | | | | | |
| | | | | Student 15 | | | | | |
| | | | | Student 16 | | | | | |

**Whole-class instruction**
The entire class is struggling with real-life connections, and most also with reflection. Even students 2, 8, and 9, although proficient, would benefit from extra metacognitive work.

**Small-group instruction**
Students 3, 5, 7, 10: Picture box; describe the picture

Students 2, 3, 10, 13, 16: Expanded form and skip counting

Students 3, 6, 7, 8, 10, 13, 16: Number line (two groups)

**Individual instruction**
Student 3 shows up as at risk in all areas of SNAP.

Extra one-to-one support is needed to ensure student 3 finds success on this learning outcome.

*Source: Chilliwack School District (2016). Used with permission.*

**FIGURE 5.16:** SNAP number sense class profile for whole-number task example.

**Small-group instruction:** Based on this class profile, some groupings are beginning to emerge. In fact, the teacher could form four small groups.

**Group 1**: Four students focusing on representation

**Group 2:** Five students working on expanded form and skip counting

**Groups 3 and 4:** Seven students split into two groups with a focus on number line skill building

We recommend a U-shaped small-group instruction table in each room to conduct these small-group lessons. We've had positive experiences with *daily 5* classroom management strategies that facilitate the opportunity to plan for small-group instruction during the regular school day (Boushey & Moser, 2014). This teaching approach creates space in the classroom for teachers to work with small groups of students while the rest of the class engages in center-type activities. Each center has a slightly different focus (for example, working with manipulatives, mathematics games, problem solving, zoom in on a section of the SNAP, and so on), and students can work independently from the teacher in each case. Teachers scaffold the daily 5 process by improving students' stamina until the students can work at each station for a substantial amount of time.

**Individual instruction:** Student 3 requires a lot of intervention to meet this learning standard, and will need over-and-above instruction from teachers to find success.

## Chapter Summary

Teachers can use the SNAP as a competency-based formative, diagnostic, or summative assessment. Using rubrics and classroom profiles, teachers can direct their instruction in targeted ways using whole-group, small-group, or individualized approaches. Teachers and students alike can track growth over time in each area, ultimately leading to more student understanding and confidence in number sense. In chapter 6 (page 95), you'll learn about how the SNAP supports response to intervention (RTI).

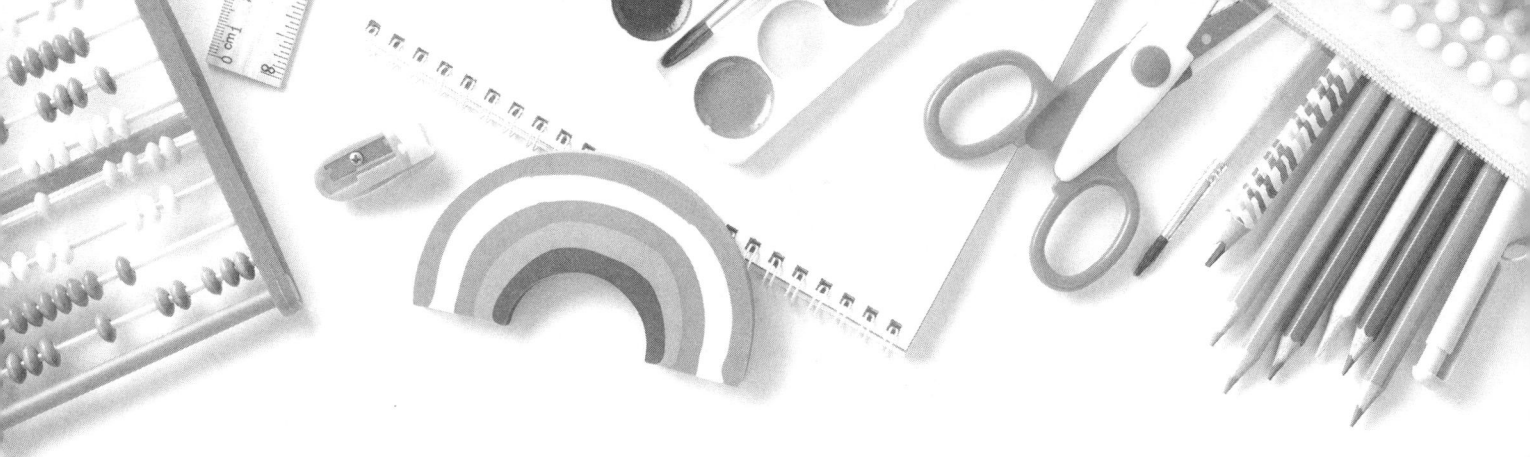

## CHAPTER 6

# Understanding How the SNAP Supports Response to Intervention

This chapter focuses on RTI and how the SNAP can support this approach in mathematics instruction. In their book *Pyramid Response to Intervention*, RTI at Work experts Austin Buffum, Mike Mattos, and Chris Weber (2009) discuss the RTI approach as:

> A new movement that shifts the responsibility for helping all students become successful from the special education teachers and curriculum to the entire staff, including special *and* regular education teachers and curriculum.... Human resources (including classroom teachers, speech and language pathologists, psychologists and social workers, special education teachers, and administrators) will be deployed in new ways to collectively assist all students.... In this unified system, assessment—universal, ongoing, and formative—assumes an increasingly important role in classrooms and schools. (pp. 2–3)

Knowing how vital universal, ongoing, and formative assessment is in the RTI framework, the SNAP provides real-time data on students' mathematical understanding, allowing teachers to plan targeted interventions and support for individual students, small groups, or the whole class. We discuss how the SNAP fits into the RTI pyramid of intervention (Buffum et al., 2009; see figure 6.1, page 96) and provide examples of how to use the SNAP formatively and summatively. We also suggest techniques for engaging SNAP activities, such as error hunts and themed templates. Using the SNAP as part of an

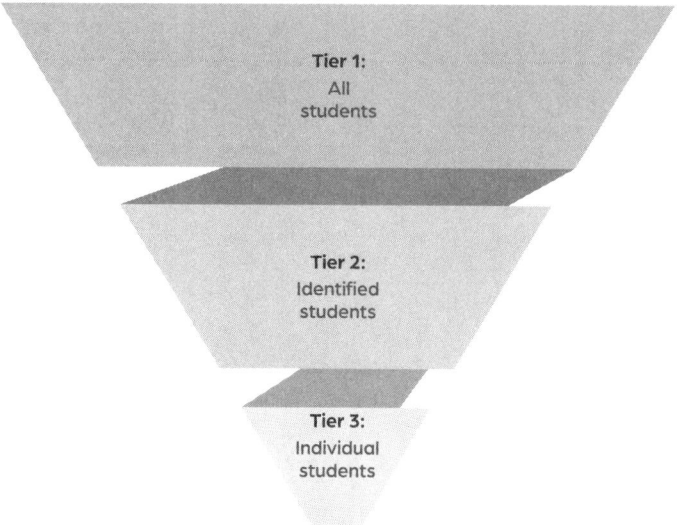

*Adapted from Buffum, Mattos, & Weber, 2012.*
**FIGURE 6.1:** RTI pyramid of intervention.

RTI approach creates environments where teachers can effectively differentiate instruction and learning thrives. Teachers can also incorporate the SNAP into a multitiered system of supports (MTSS) approach. As coauthors Tom Hierck and Chris Weber (2023) suggest, "Multitiered systems of support (MTSS) are about using the knowledge, skills, and attributes of all members of a learning organization to positively impact the life chances of all students. It is common sense in action" (p. 1). The SNAP aligns with the MTSS approach by identifying gaps to assist staff in providing targeted support to struggling students. This chapter discusses how the SNAP works with RTI, with specific ideas for each of the three tiers.

## Supporting RTI

An RTI approach is powerful in any teacher's tool kit. Educators who are well-versed in planning for high-quality, differentiated instruction through the lenses of data-driven decisions, collective responsibility, and targeted interventions create environments where learning thrives. In their book *Taking Action: A Handbook for RTI at Work,* coauthors Austin Buffum, Mike Mattos, and Janet Malone (2018) explain:

> Equally important, we know that a successful system of interventions must be built on a highly effective core instructional program, as interventions cannot make up for a toxic school culture, low student expectations, and poor initial instruction. Fortunately, our profession has a near unanimous agreement on how to best structure a school to ensure student and adult learning. (p. 3)

Buffum and colleagues (2018) add, "Interventions are most effective when they target a student's specific learning needs. This requires assessment data that can identify the specific standard, learning target, skill, or behavior that a student lacks" (p. 8).

The data need to be *valuable*—meaning the data are accurate and reflect the learning that has occurred or is missing in a student's learning profile. Teachers, working in collaborative teams and using the SNAP as a common assessment, work together, use the data, and collectively support students with effective intervention and support.

The SNAP offers a window into students' thinking by providing educators with clear and concise demonstrations of student mathematical understanding. The collected data from the SNAP point directly to the varied foundational aspects that every student must know. For example, if a student cannot place a number correctly on a number line, that is a gap in learning and needs immediate attention. Whether it is a seven-year-old struggling to position 15 on a line between 0 and 20; or a twelve-year-old who does not understand that ½ or 0.5 is midway between 0 and 1, these data provide valuable and insightful information.

Teachers can formatively preassess an entire class in fifteen minutes using the one-page SNAP. Then, based on the class results, the teacher creates a road map to scaffold the next steps, which may include additional support (Tier 2) or intense instruction for a deficit in foundational skills (Tier 3; see figure 6.2, page 98). If a particular concept is difficult for most students, the teacher should teach this concept to the entire class. This teaching is known as *Tier 1* (or *core*) *instruction* and will benefit all students. Buffum and colleagues (2009) write:

> **The most important step a school can take to improve its core program is differentiating instruction and small-group activities. Schools must first ensure that exceptional and committed teachers are delivering research-based core programs as intended and using classwide formative assessment data to identify emerging classwide areas of need. A Tier 1 curriculum must be prioritized so that students have ample opportunity to master power standards. (p. 74)**

Suppose there is a concept only some students need help with. In that case, the data support a small-group instruction and intervention approach—with dedicated time from the classroom teacher, a learning support teacher, or other resources, or *Tier 2* support (Buffum, et al., 2009):

> **Because students struggle at school due to a variety of causes, including student effort, Tier 2 interventions must be designed to meet the needs of both failed learners (students who failed to learn) and intentional nonlearners (students who failed to try). Their needs are markedly different. A successful Tier 1 program should meet the needs of at least 75% of the student body, and an effective Tier 2 supplemental level will meet the needs of at least 15% more. (p. 89)**

| |
|---|
| **Tier 1 (All Students)** |
| 1. Use a formative assessment or screening tool prior to new units.
2. Employ a teaching tool for whole-class, high-quality instruction.
    - Provide group work.
    - Provide partner work.
    - Provide individual work.
3. Differentiate for students to assist with intervention.
    - Select developmentally appropriate doorway questions.
4. Mix up strategies to enhance engagement.
    - Students solve one another's word problems.
    - Students go on an error hunt to find mistakes in the completed SNAP. |
| **Tier 2 (Identified Students)** |
| 1. Use a diagnostic tool.
    - Why is a student struggling?
    - What foundational concepts are students missing?
2. Target the intervention
    - Use small-group instruction at correct level. |
| **Tier 3 (Individual Students)** |
| 1. Provide individualized intensive programming and intervention.
    - Deign specific IEPs or speech-language pathology.
    - Focus on foundational outcomes at the developmental level.
    - Scaffold learning.
    - Conduct one-to-one instruction and support.
2. Connect Tier 3 instruction to classroom learning environment. |

**FIGURE 6.2:** RTI tiers and the SNAP connections.

Finally, when only one or two students are unsuccessful in the concepts a SNAP identifies, the intervention is individualized and targeted for the student or students to find success, which is referred to as *Tier 3* support. Buffum and colleagues (2009) state:

> Some students who have received Tier 1 core instruction and Tier 2 supplemental interventions will continue to have difficulty learning. . . . They need instruction that is even more explicit and intensive, and even more targeted and tailored to their individual needs. Interventions that provide such instruction are known as *Tier 3 or intensive*. (p. 100)

According to education academic John Hattie (2023) in *Visible Learning: The Sequel*, RTI has an effect size of 1.09 on student achievement. This is significant, as an effect size of 0.3 indicates low impact, effect sizes of 0.4 to 0.6 indicate medium impact, and anything over 0.7 indicates a high impact on student learning. Clearly, RTI, with its focus on tiers of instruction, is an effective model that provides structures for all students to succeed.

## Supporting Tier 1

Teachers use formative assessments to informally assess student understanding of concepts. In addition, they use formative assessments to plan upcoming lessons and units, ensuring their Tier 1 instruction (learning for all students) is on point and appropriate for the whole class. Conversely, teachers use summative assessments to not only identify and evaluate student learning primarily for reporting purposes but also for opportunities to reteach essential learning outcomes. The SNAP is equally effective for both formative and summative assessment purposes. For example, the SNAP is an excellent formative preassessment and screening tool for teachers to use before teaching new units or concepts. Teachers quickly identify missing concepts when using the SNAP with individual students, providing real-time data and trends for teachers to plan and provide immediate feedback to each student. Teachers can use a class profile sheet (see figure 6.3) to document the themes (or patterns) they glean from student work. Once a teacher builds a class profile, the next steps for teaching are clear. We discussed this process more in chapter 5 (page 71).

## SNAP Number Sense Class Profile

**Teacher:** _____ **Year:** _____ **Doorway question:** _____

**Instructions:** For each doorway question, add each student's name under *not evident* (the student isn't *proficient*) or proficient. Next, make notes about whole-class, small-group, and individual instruction.

| Conceptual understanding<br>*Picture box and description picture* | Procedural fluency<br>*Expanded form and counting forward and backward* | Strategic competence<br>*Real-life connection and three equations* | Adaptive reasoning<br>*Number line* | Productive disposition<br>*Reflection* |
|---|---|---|---|---|
| Not evident / Proficient | Not evident / Proficient | Not evident / Proficient | Not evident / Proficient | Not evident / Proficient |

| Whole-class instruction | Small-group instruction | Individual instruction |
|---|---|---|
| | | |

*Source: Chilliwack School District (2016). Used with permission.*

**FIGURE 6.3:** Classroom profile sheet for the SNAP template.

*Visit go.SolutionTree.com/mathematics for a free reproducible version of this figure.*

If there are concepts in a lesson the whole class needs help with, then the teaching plan must cater to everyone through whole-class instruction. As with all effective instruction, differentiation must be present in the teacher's lesson design and delivery. For example, suppose the class concept is place value. Teachers can provide students who need stretch goals (they have mastered the concept) a more difficult doorway question (in tenths), and teachers can give those students who need a more specific access point to the work (they need scaffolding to attain proficiency) a more accessible number to start with (in tens; see figure 6.3).

How teachers organize learning activities varies. The SNAP complements multiple teaching styles and provides student access to learning. Teachers can use the SNAP with the whole class as a demonstration lesson or as an engaging and challenging group activity with students working together to find solutions and examples of doorway concepts. The opportunity to partner with students provides social interaction with back-and-forth negotiation and conversation, leading to enhanced understanding and lower anxiety. Mathematics is a content area where anxiety and a fixed mindset originate very early in a learner's school career. Psychologist and motivation and mindset researcher Carol S. Dweck (2015) expounds:

> Such children hold an implicit belief that intelligence is innate and fixed, making striving to learn seem far less important than being (or looking) smart. This belief also makes them see challenges, mistakes and even the need to exert effort as threats to their ego rather than as opportunities to improve. (p. 12)

A *growth mindset* in adult learning (where adults work in pairs and groups) informs teaching, which can translate into a growth mindset in student learning as well. If educators use a growth mindset to support adult learning, they must provide at least the same to support students by allowing them to work in pairs or small groups, which reduces anxiety and helps create a supportive and growth mindset in all students.

It is necessary to keep things new and exciting for students, and although seeing a familiar template brings down anxiety, it can also bring on a "not this again" response. To assist with the need to interweave curricular concepts to ensure deep and flexible learning and balance the boredom factor, we have observed teachers adopt the following effective techniques when employing the SNAP.

1. **Error hunt:** The teacher completes the SNAP on the whiteboard with deliberate errors. Students find and correct the mistakes. Students enjoy catching errors and pushing the teacher's thinking.

2. **Students create word problems:** With this technique, students create word problems and solve them using the real-world examples the SNAP provides. This activity is enjoyable for the students because they make meaning for one another.

3. **Zoom in on the SNAP:** Instead of using the entire template, teachers zoom in on one aspect of the SNAP and create a lesson and activity on that competency. For instance, a teacher may zoom in on just the number line and begin a number talk lesson on just number lines.

Grade 4 Vedder Elementary School teachers Andy Fast and Paul Wojcik designed some creative and unique SNAP templates to both zoom in on the SNAP and provide novelty to prevent the SNAP from becoming just another worksheet. For counting forward, the teacher asks students each to place a number on the bottom stair and "walk upward" with their number and vice versa for counting backward (see figure 6.4).

**FIGURE 6.4:** Counting forward and backward.

4. **Artistic and themed SNAP templates:** Teachers can adapt the SNAP to themed templates that create extra student interest. Students use these themed templates to develop real-life applications and connect their numeracy learning to cross-curricular topics. Figure 6.5 (page 102) displays three examples of themed templates.

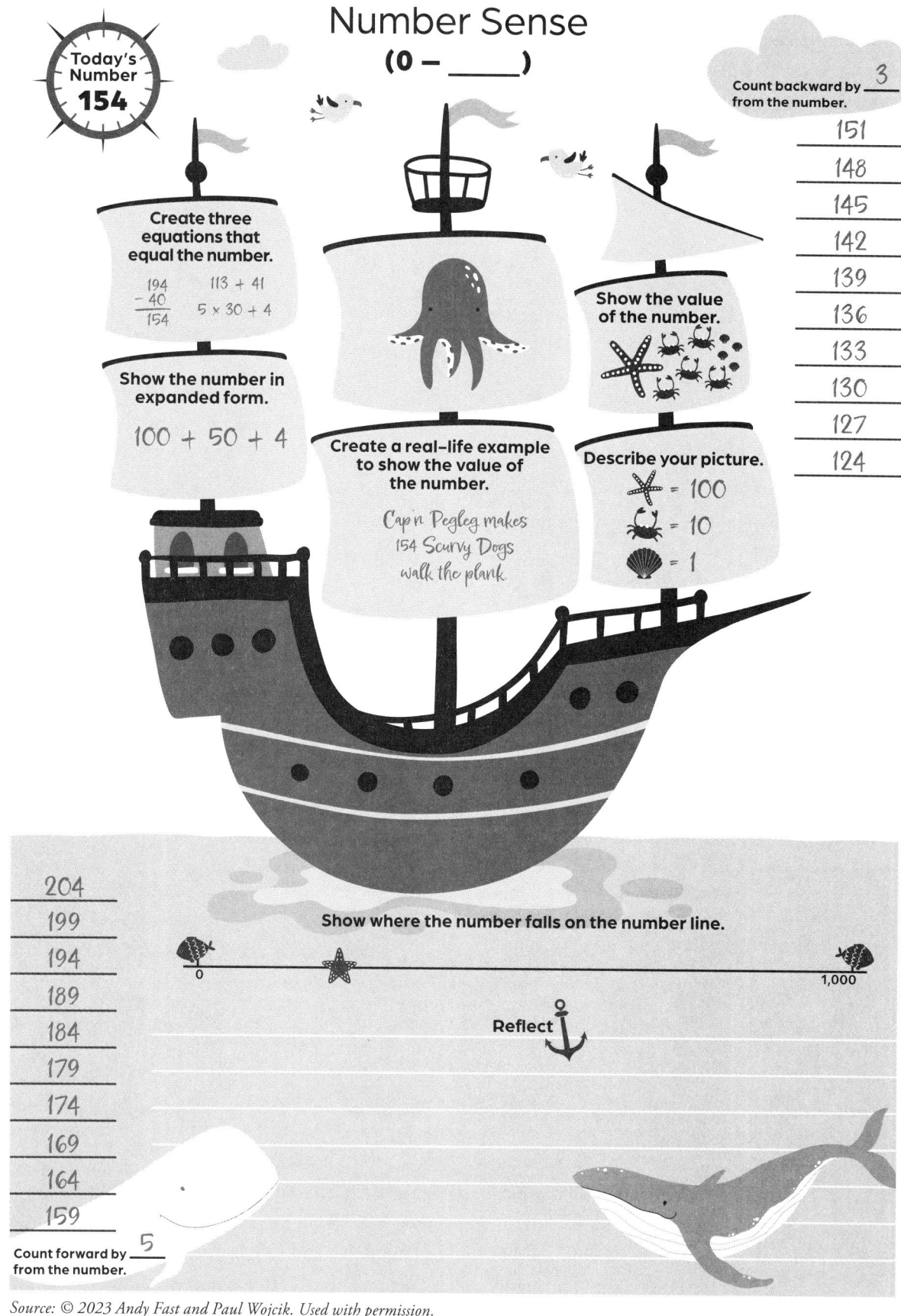

**FIGURE 6.5:** Sample artistic and themed SNAP templates

## SNAP Number Sense

**0 – 10,000**

Today's Number

Show the number in expanded form.

Create three equations that equal the number.

Create a real-life example to show the value of the number.

Count backward by _____ from the number.

Show the value of the number.

Count forward by _____ from the number.

Describe your picture.

Show where the number falls on the number line.

**Reflect**

*continued →*

# SNAP
## Number Sense
### 0 – 10,000

**Today's Number**

**Show the number in expanded form.**

**Create three equations that equal the number.**

**Count forward by ____ from the number.**

**Count backward by ____ from the number.**

Show where the number falls on the number line.

0 — 10,000

**Show the value of the number.**

**Create a real-life example to show the value of the number.**

**Describe your picture.**

**Reflect**

*Visit **go.SolutionTree.com/mathematics** for a free reproducible version of this figure.*

5. **Guess my number:** The teacher creates a starting point by partially filling in one or two sections of the SNAP, but leaving the middle oval blank. Then, students must understand and complete the rest of the SNAP to figure out the number in the middle. Students do this activity with partners, where students each fill in parts of their own SNAP and then the partners exchange SNAPs to complete and solve.

6. **Mix and mingle:** Divide the class into groups of three or four. Each group is responsible for filling in one SNAP together. The teacher chooses one of two possible numbers to whisper to each group. The group cannot write the number in the middle oval. Instead, each group works together to fill in the SNAP based on the assigned number from the teacher. Once the groups finish, the teacher says, "Mingle!" Students must then find another group whose number matches theirs by comparing their completed SNAPs. All groups should skip count forward and backward by the same amounts.

7. **Student choice:** The teacher provides students a choice of three different numbers written on the board, and students each write one of the numbers in the oval of their own SNAP. The teacher offers easy, medium, and challenging numbers to differentiate the practice. Alternatively, students can choose any number they want (within the grade-level parameters).

## Supporting Tier 2

When the evidence teachers collect formally or informally reveals student confusion on the Tier 1 formative assessments, there should be alarm bells sounding. If students are missing foundational learning, they risk falling behind the pace of learning, which can create anxiety, confusion, and loss of confidence. In turn, student misbehaviors may increase—ranging from anger to avoidance. Educators and coauthors Anne Fitzsimons Hughes and Beatrice Adera (2006) reveal that meaningful, relevant, and emotionally and intellectually engaging classroom instruction minimizes inappropriate student behaviors. Because the SNAP looks the same at all complexity levels, teachers can use the same template for each student, regardless of skill or level of understanding. Teachers can also plan to mitigate the impact of misbehavior through familiarity and clarity, which reduces the intensity of the students' struggle. Researchers Robert J. Marzano (2003) and Brandi Simonsen, Sarah Fairbanks, Amy Briesch, Diane Myers, and George Sugai (2008) posit that teachers must be clear and consistent about what they expect students to do (as well as teach them how to do it); teacher clarity is a key to student achievement. The earlier teachers intervene, the smaller the learning gaps and the greater the likelihood of closing those gaps. University researchers Lina Shanley, Ben Clarke, Christian T. Doabler, Evangeline Kurtz-Nelson, and Hank Fien (2017) reveal that adults often introduce early number skills to young learners prior to school entry, and those skills link to future mathematics achievement. When they enter kindergarten, young learners have a range of preschool experiences and widely variable academic skills, so a focus on early number skills in kindergarten is crucial to closing achievement gaps.

The teacher needs to answer the following questions to plan for students with learning gaps.

- Why is the student struggling?
- What foundational concepts is the student missing?

Based on where students each are struggling on grade-level concepts, using the SNAP template will offer insight into where the teacher goes next. For instance, if the grade-level concept is working with number sense 0–1,000, and a student is unable to place 157 on a number line, the teacher can roll back the learning outcome a progression at a time until the teacher finds the student's current level of performance. For instance, if 157 is too complex—can the student place 57 (0–100) on a number line? What about 15 (0–20)? How about 5 (0–10)? Working backward (or forward) with a struggling student will identify the student's current performance level and create a starting point for just-right instruction. The following scenario may be familiar.

> **The resource teacher is supporting a grade 5 student with autism named Max. His teachers and the school-based team are still learning the best ways to support him. Sometimes, Max seems able to fly effortlessly through his mathematics work; other times, he is confused. The resource teacher used the SNAP template as a diagnostic assessment to work through the curriculum to narrow in on Max's performance level. Instead of working backward, she started with the grade 1 curriculum (number sense to 20). Grade 1 curricula was simple for him. Next, she moved to the grade 2 curriculum (number sense to 100) and simple addition. Again, grade 2 curricula was very easy for Max. She then moved to grade 3 curricula. The SNAP templates showed Max was lost when the numbers jumped to the 1,000s. The team could confidently create an individualized learning plan that met Max where he was, which was solidly landing in grade 3 learning outcomes.**

Using the class profile sheet in figure 6.3 (page 99), teachers can identify students who need support in specific common areas. Teachers can then group students with similar needs and create targeted interventions for small-group instruction. Teachers will sometimes use a strategy where they focus in on one aspect of the SNAP for targeted instruction.

Much like focusing on the individual skill of passing the puck in hockey, there are times when teachers would like to focus on a specific area of the SNAP separate from the other skills. Using the whole template can sometimes overwhelm students or they become complacent with the lack of uniqueness. In addition, when using the classroom profile, there are times when a teacher wants to zoom in on a particular skill with the whole class or a small group. Figure 6.6 shows SNAP templates that zoom in on only one section at a time.

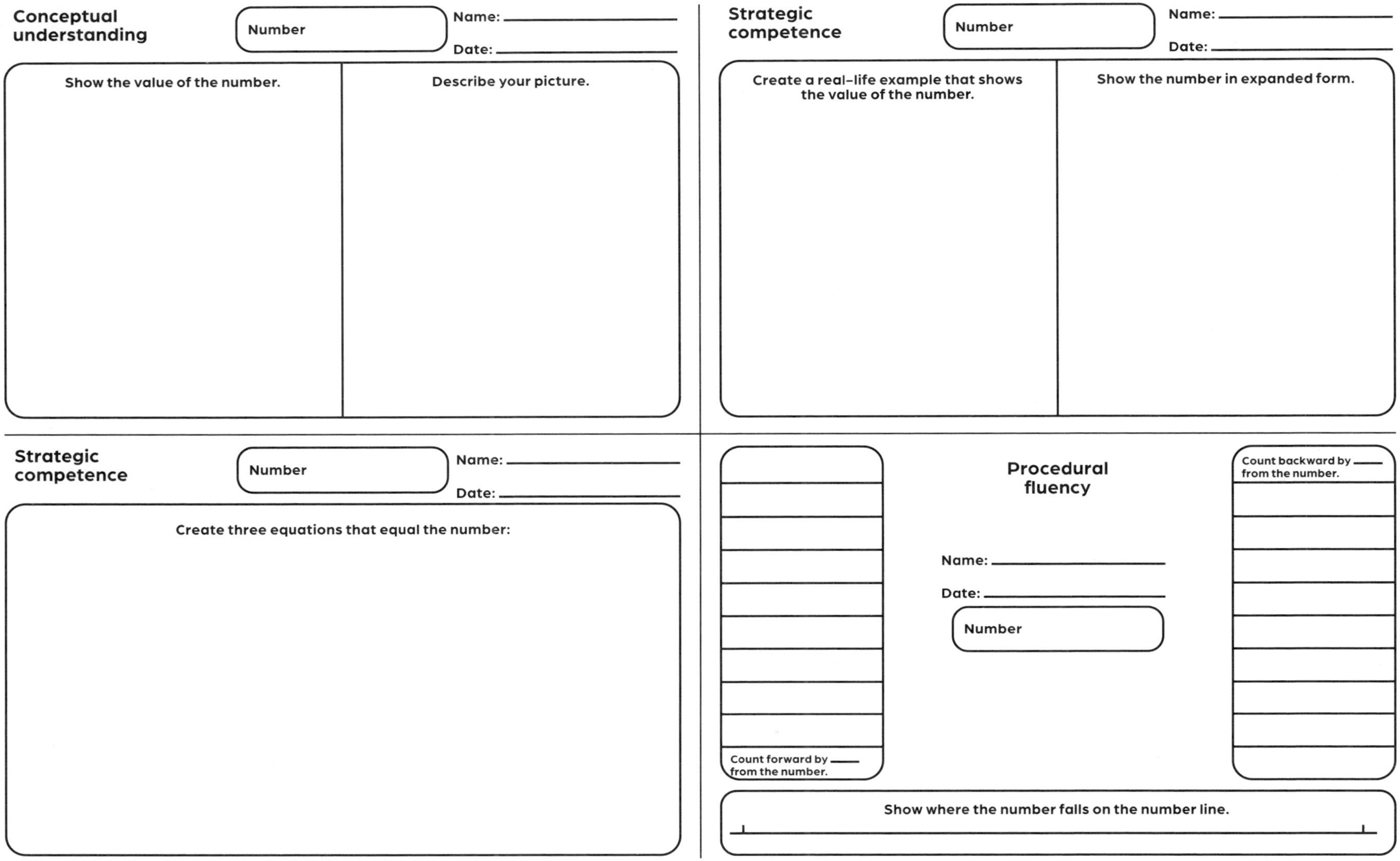

*Understanding How the SNAP Supports Response to Intervention* 107

### Conceptual understanding

Number ⎯⎯⎯ | Name: ⎯⎯⎯ Date: ⎯⎯⎯

- Show the value of the number.
- Describe your picture.

### Strategic competence

Number ⎯⎯⎯ | Name: ⎯⎯⎯ Date: ⎯⎯⎯

- Create a real-life example that shows the value of the number.
- Show the number in expanded form.

### Strategic competence

Number ⎯⎯⎯ | Name: ⎯⎯⎯ Date: ⎯⎯⎯

- Create three equations that equal the number:

### Procedural fluency

Number ⎯⎯⎯ | Name: ⎯⎯⎯ Date: ⎯⎯⎯

- Count backward by ⎯⎯ from the number.
- Count forward by ⎯⎯ from the number.
- Show where the number falls on the number line.

*Source: Chilliwack School District (2016). Used with permission.*

**FIGURE 6.6:** Zooming in on one section at a time in the SNAP templates.

*Visit* **go.SolutionTree.com/mathematics** *for a free reproducible version of this figure.*

## Supporting Tier 3

Max's story fits all too well in the typical post-pandemic classroom; too many students have significant gaps in their learning. Max's learning support would undoubtedly fall into a Tier 2 intervention in today's classroom, as his learning needs align with others in the classroom. Teachers can deliver support via small-group instruction inside or outside the classroom—as per the definition of Tier 2 (Buffum et al., 2009).

However, if his needs were profound, Max may have needed more individualized support. For example, suppose Max was still working on subitizing, counting, and beginning number sense concepts in his grade 7 class. In that case, receiving teacher support directly in the classroom might not work well for him. As a middle school student, Max may be uncomfortable working at levels well below his peers. In this scenario, the teacher may need to move the intervention to one-to-one instruction outside the classroom environment. As long as the intervention teacher uses completing the SNAP as the goal for demonstrations of learning, the integrity of the learning remains intact, and we can be confident that Max would transition to classroom learning once he catches up. However, let us be clear: a student (like Max in our example) would not receive Tier 3 support instead of Tier 1 (or core) instruction, but as support in addition to the Tier 1 instruction and Tier 2 support. The more unique the challenge or gap, the more individualized the solution. The SNAP tool provides clarity on where the intervention begins for each student and how it can continue to ensure all students reach proficiency on the desired concepts.

## Chapter Summary

The SNAP and both RTI and MTSS are complementary, robust tools for educators to support students in achieving academic success. The SNAP offers a comprehensive approach to assessment, providing teachers with real-time data to inform instructional decisions and interventions. RTI is an evidence-based approach to identifying and supporting students with learning and behavioral challenges, allowing educators to intervene early and improve student outcomes. Data are the engine that drives a successful RTI or MTSS approach, and the SNAP is the tool that delineates the data. By combining these tools, educators can create a culture of data-driven decision making that supports student growth and achievement. While implementing these approaches may require an initial investment in time and resources, the benefits to students are worthwhile. In chapter 7, you'll see how the SNAP relates to intervention approaches, competency-based education, and a school or district implementation.

# CHAPTER 7

# Looking at School and District SNAP Implementation

We cannot overstate the importance of assessment in education. Well-developed assessments enable teachers to tailor their instruction to meet the specific needs of students and help identify areas where students may be struggling. In addition, by implementing *common assessments* (or assessments multiple teachers share), educators can ensure all students receive a high-quality education, which research proves boosts student achievement. A report from the Organisation for Economic Co-operation and Development (OECD, 2012) states, "The highest performing education systems are those that combine equity with quality. They give all children opportunities for a good quality education" (p. 3). Marzano (2003) reveals that in classrooms where teachers identified as *most effective* lead, students post achievement gains of 53 percentage points over the course of one academic year, whereas in classrooms where teachers identified as *least effective* lead, student achievement gains averaged 14 percentage points over the same time frame.

The SNAP can significantly improve the education system. Educators created the SNAP to use in regular classrooms (not the standardized assessments schools usually use). By design, the SNAP is more personalized and effective for differentiated instruction. Large-scale assessments can be difficult for teachers to find value in because they are often impersonal, culturally insensitive, and not tailored to individual student needs. In contrast, the SNAP is a more welcoming alternative to traditional state or provincial evaluations. Students and teachers can develop familiarity with use, and the SNAP produces reliable and valid results (Savage, 2021). The familiarity and acceptance of the SNAP can be very helpful in implementing a systemwide change process, which can otherwise be difficult due to staff resistance to change, lack of resources, and competing priorities. Michael Fullan (2007), an author known for his expertise on educational reform, states:

> A large part of the problem of educational change may be less a question of dogmatic resistance and bad intentions (although there is certainly some of both) and more a question of the difficulties related to planning and coordinating a multilevel social process involving thousands of people. (p. 84)

In this chapter, we discuss how to implement the SNAP at a school or district level, including data analysis and monitoring, cut scores, transparency and accountability, training, and school leadership.

## Implementation at the School or District Level

Implementing the SNAP schoolwide or districtwide has many practical benefits. For example, a successful implementation ensures all teachers, administrators, and district decision-makers are on the same page when assessing and planning for student achievement, which promotes better communication and collaboration across departments. In addition, the shared data promote collective responsibility, which leads to improved transparency and accountability throughout the system—from the classroom and teacher team meetings to the principal's office and boardroom.

The first step to systemwide implementation is to establish the non-negotiable skills students must learn at each grade level. In the Chilliwack School District, the leaders asked teachers to identify two foundational mathematical skills students should learn at each level, from grades 2–7. A committee of teachers identified the following number sense and operation outcomes based on the curriculum for each grade level (see figure 7.1). Please note that this book is focused on SNAP–Number Sense and corresponds with the number sense goals in this table. SNAP–Operations is in planning, and we have provided both outcomes for system-wide implementation in this example.

| Grade | Number Sense | Operations (Sample Operations) |
|---|---|---|
| 2 | • Number concepts to 100<br>• Any two-digit number | Addition of two-digit numbers without regrouping<br>24 + 33<br>51 + 17 |
| 3 | • Number concepts to 1,000<br>• Any three-digit number | Subtraction of three-digit numbers with regrouping<br>427 − 152<br>754 − 226 |
| 4 | • Number concepts to 10,000<br>• Any four-digit number | Multiplication of a one-digit number by a three-digit number<br>4 × 326<br>7 × 142 |

| 5 | • Number concepts to 1,000,000<br>• Any six-digit number | Division of a three-digit number by a one-digit number with remainder<br>625 ÷ 3<br>291 ÷ 4 |
|---|---|---|
| 6 | • Number concepts thousandths to billions<br>• Any decimal to the hundredths | Division of four-digit number to hundredths by one-digit whole number (quotient should not exceed thousandths)<br>45.34 ÷ 5<br>71.76 ÷ 3 |
| 7 | • Integer concepts<br>• Any two-digit negative whole number | Percentage of calculations (Find the percent of a number. Answers should be in the tenths or hundredths.)<br>16 percent of 85<br>47 percent of 42 |

*Source: Chilliwack School District (2016). Used with permission.*
**FIGURE 7.1:** Number sense and operations skills in grades 2–7.

These outcomes became the non-negotiable curriculum in the Chilliwack School District. We recommend following a similar process in your school or district. Create an assessment team of educators to review your state, provincial, or district learning standards. Line up these standards with the SNAP, and prioritize one or two foundational outcomes. For these outcomes, all teachers in the district use the SNAP to assess student proficiency. Teachers and students have the entire year to meet these outcomes. Students who are proficient in November do not need to continue to prove their skills. Only those who are not proficient must continue to receive support and interventions to achieve grade-level performance. Tracking the SNAP data has been essential to identifying these students.

## Data Analysis and Monitoring

The Chilliwack School District has tracked SNAP data since 2017. The information is valuable for highlighting areas of strength and areas where staff and students require extra resources in mathematics instruction. Currently, the district collects data through teacher input about the roughly 6,000 students in grades 2–7. In previous chapters, we discuss how teachers use the students' results at the classroom level for formative assessment purposes, but district leaders also lean into these data. For instance, there is a noticeable drop in achievement during the transition from grade 5 to grade 6. This dip in achievement also corresponds to when students move from elementary school to middle school. To mitigate this achievement concern, district leaders responded by adding extra resources and training at the middle school level.

Additionally, district educators can use the SNAP to triangulate data collection between classroom-based assessment (report cards) and the Foundation Skills Assessment, a standardized test in British Columbia, Canada, for all grades 4 and 7 students. A study shows a moderately strong correlation of approximately 0.6 between these three assessments (Savage, 2021). A correlation of 1.0 is an exact match, 0.8 is very strong, and 0.6 is moderately strong. A correlation of 0.0 would represent no connection between variables. Information and analysis help to create confidence in the teachers' work in the classroom, as well as provide evidence all three processes (classroom-based, large-scale standardized provincial, and competency-based) have educational value.

School systems should pay attention to large-scale assessments and how they relate to local data trends. For instance, 2003 mathematics achievement for fifteen-year-olds worsened for Canadians and Americans based on Programme for International Student Assessment (PISA) scores. Canada's achievement dropped steadily from 532 in 2003 to 512 in 2018. The United States drop-off is not quite as dramatic (483 to 478), but its average score is well below the Canadians (OECD, 2004, 2019).

## Cut Scores

When teachers use the SNAP for summative purposes (report cards) and district leaders need the same data for system-performance checks, they can calculate an overall score for the SNAP. Each of the five strands of the assessment (conceptual understanding, procedural fluency, strategic competence, adaptive reasoning, and productive disposition) have a maximum of four points, totaling a score of 20. Here are the recommended cut scores.

- **0–7:** Emerging
- **8–12:** Developing
- **13–16:** Applying
- **17–20:** Extending

By using these cut scores, teachers can quickly assign an overall summative grade for each completed SNAP. These scores translate easily for school or district data collection and analysis.

## Transparency and Accountability

The way educators teach mathematics in schools has undergone significant changes since 2000. Instead of traditional teaching methods, educators today emphasize competency-based learning, as well as require students to showcase skills such as adaptive reasoning and a positive attitude toward problem solving in mathematics. This shift in teaching approaches has left many parents feeling uncertain and perplexed about what exactly these changes entail, and how the changes impact their children's education. Regular use of the SNAP can help eliminate many parents' questions and concerns. We have repeatedly seen parents (who at first are confused about how their child is progressing in mathematics) be won over when teachers present them the one-page demonstration of their child's knowledge on a SNAP. A common question parents ask is, "Why didn't my child receive an A or an exceeding expectations grade?" However, when they see the level of understanding for themselves, and the teacher (or even better, the student) explains what the student needs to do to bolster learning, we find that parents are on board and supportive.

In a similar manner, when student learning struggles are clear for teachers and parents, IEPs can be purposeful, transparent, and helpful. The SNAP brings it all to the table by providing a window into student thinking that helps teachers create visible, measurable, and scaffolded interventions as needed. The clearer educational outcomes are for all involved, the more successful the learning organization.

When strategies work in the classroom for all learners, decision makers at the board of education take note. The SNAP provides evidence-based assessments clear and simple enough for teachers to use daily, parents to understand and support, and decision makers to trust because it supports hot-spot areas in a school or district. Based on student results over time, these adults can respond with confidence and deliver extra resources and training to areas of most student need.

## Training

Implementing the SNAP districtwide is a process that requires attention from school and district leadership to succeed; students and teachers must look at mathematics differently while using the SNAP. Most teachers need to gain experience understanding mathematics at a conceptual level. Many students struggle to go beyond rote calculations when interacting with equations and mathematics problems. Connecting mathematics to real-life experiences is challenging and requires teachers to transform their thinking and teaching techniques, so they need more support from district leaders.

We suggest the district leadership provide training that allows teachers to work together to plan a lesson introducing the SNAP as a formative assessment, which students then work on. After the lesson, leaders must provide facilitated time for the same group of teachers to review student responses and collaboratively anchor student performance to the curricular outcomes. Once teachers see the SNAP in action and begin teaching and assessing in new ways, they understand how the SNAP will improve teaching and learning in the classroom.

We find what works best is a structured in-service training program, where teachers have three opportunities to learn side by side with a mathematics coach throughout the year. The workshops are ninety minutes long and follow a structure that balances understanding the guiding principles of the assessment as well as time in the classroom with students. A typical training plan might look like the following.

> **Day one:** Workshop with a grade level of teachers (for example, all grade 4) to understand the guiding principles of the SNAP.
>
> **Day two:** Introduce the SNAP to a class of students and conduct a formative SNAP. Review and anchor the SNAP as a facilitated activity.
>
> **Day three:** Create a class profile of the students' needs with intervention strategies for the whole class (Tier 1) and small groups (Tier 2).

## School Leadership

Principals and vice principals are key players in any schoolwide implementation. With the SNAP implementation at the Chilliwack School District, school leaders provided classroom teachers and *non-enrolling staff* (for example, learning assistance teachers, librarians, and school psychologists) with team time (biweekly collaboration time) to discuss the new assessment and how to use it with students. Teachers then analyzed data sets, and students who needed further instruction received an intervention lesson. Non-enrolling staff also support small-group interventions.

When staff are grappling with an innovative approach, district leaders should seize any opportunity to highlight and show off staff's successes, and also examine failures to help solidify implementation. Several Chilliwack schools hosted guests from Quebec, Canada, and Illinois interested in learning more about both RTI and the SNAP. Not only did the visitors gain a new understanding but also the host schools reflected deeply on their learning journeys, creating more buy-in and a sense of pride in their innovative work.

As with any new initiative, leaders must understand the realities of an *implementation dip*. Teacher, entrepreneur, and author Seth Godin (2007), author of *The Dip*, states, "If it is worth doing, there's probably a dip" (p. 21). Godin (2007) is not wrong; most initiatives will suffer from early setbacks and things may get worse before they improve. The SNAP implementation is no different; adults and students throughout the system must learn how to use the SNAP before the organization will realize the assessment's benefits.

## Chapter Summary

Implementing a new initiative like the SNAP in schools and districts can be a dynamic, challenging, and extremely worthwhile endeavor. Having a common assessment like the SNAP promotes coherence in school and district student achievement. Using the SNAP data is useful in guiding learning decisions, including how best to provide interventions in classrooms, schools, and districts. Transparency and accountability are imperative at all levels, and involving parents (who may be mathematics wounded) in mathematics conversations using the SNAP is extremely advantageous. It is essential to establish a clear plan and training cycle when implementing the SNAP in schools, and support must include district staff and school leaders, with special emphasis on providing time for teacher collaboration.

# EPILOGUE

# Final Word

Understanding the foundations of number sense is critical for student development of mathematical skills. Equally, the lack of development of number sense can exacerbate challenges students experience early and further a belief that mathematics might be beyond their grasp. In preparing this manuscript, we heard from colleagues who could connect students' early acquisition of number sense with their future success. Often, these colleagues framed this message with a pessimistic view that we paraphrase: *The amount of mathematics knowledge students came into kindergarten with showed a big difference in the success of students' short-term mathematics development.*

In other words, if students didn't have number sense and teachers didn't have a tool ready to help delineate their challenges, students' struggles not only continued but also grew into mathematics phobia and reluctance! The SNAP offers a lifeline to students and assists teachers in determining students' challenges; the SNAP readily delineates the next steps for them. And it's not only about the early years but also about long-term implications. As coauthors Jemmy Louange and Jack Bana (2010) point out:

> The study showed that there is quite a strong correlation between the number sense and problem solving proficiency of year 7 students. The evidence points towards a relationship in which problem solving performance depends on number sense proficiency more than the latter depending on the former. (p. 365)

Researchers Nancy Jordan, Joseph Glutting, and Chaitanya Ramineni (2010) also strongly suggest that "weaknesses in intermediate symbolic number sense, or number competencies related to counting, number relationships, and basic operations, underlie most mathematics learning difficulties (e.g., Geary et al., 2007; Gersten, Jordan, & Flojo, 2005; Landerl, Bevan, & Butterworth, 2004)." An easy-to-understand and deeply effective tool goes a long way in addressing the challenges of student assets and dispositions while supporting the occasional mathematics-phobic educator. The SNAP is the tool, and the practice of numerous educators validates its credibility (Savage, 2021).

Throughout this text, we explain the origins of the SNAP tool and its alignment with the five strands of mathematical proficiency the National Research Council (2001) outlines. Each skill in the SNAP tool aligns with a particular strand, and we include how to address each skill from the perspectives of various early years educators. We examine how the SNAP tool is an essential part of an effective assessment protocol, including the ease with which teachers can create rubrics and how the SNAP can assist with the transition to competency-based learning. The SNAP also brilliantly aligns with RTI and MTSS frameworks; evidence teachers gather from the SNAP makes planning for all students' success in mathematics very straightforward. Finally, we outline how a school and district can move forward with the SNAP implementation.

Mathematics is an essential skill for every adult to experience success in their lives. We equally believe all students love mathematics long before they hear the term; they enjoy counting, sorting, and comparing. Nurturing all students' enjoyment in both the short term (building the foundation) and the long term (comfort and competency with numbers as an adult) is achievable! It begins with educators' first interaction and desire to gather quality evidence to make plans for a life of mathematics proficiency (not make excuses and further a fear of numbers). We trust the work we describe throughout this text allows you and your colleagues to nurture the enjoyment, accessibility, and wonder of mathematics!

# APPENDIX A

# SNAP Templates, Rubrics, and Classroom Profiles

We advise you begin by using the detailed SNAP template appropriate for your grade level. As you and your students become familiar with the SNAP, explore the castle, pirate ship, or skipping rope templates (figure 6.5, page 102), or create your own templates to promote a sense of novelty.

We have also provided open-ended and blank SNAP templates where you can add your own sets of questions or tasks for your students, which can also be suitable for use at higher grades.

Finally, the appendix includes reproducibles for the SNAP Number Sense Rubric and SNAP Number Sense Class Profile.

# Kindergarten Number Sense SNAP

Name: _____

| Build the number. One or two more or less | Represent and subitize the number in three ways. |
|---|---|
| + 1 =     + 2 = <br> −1 =     −2 = | |
| **Decompose the number in three ways.** | **Benchmarks of five and ten** — Tell me where you would be able to see _____ of something in your world. |

Circle the number on the number path (or line).

# Grade 1 Number Sense SNAP

Name: _____

**Draw a picture to represent the number.**

**Draw the number on the ten frames.**

Count backward from _____.

**Show the number that is one more or less or two more or less.**

___, ___, ☐, ___, ___

**Decompose the number to make three equations.**

**Talk about a real-life example that shows the value of the number.**

Count forward by 1, 2, 5, or 10 from the number.

**Show the quantity on the number path (or line).**

0 ——————————————— 20

## SNAP for 0–100

# SNAP
## Number Sense
### 0–100

Name: _____

Date: _____

Count forward by _____ from the number.

Count backward by _____ from the number.

Show the value of the number.

Describe your picture.

Show the number in expanded form.

Create three equations that equal the number.

Create a real-life example that shows the value of the number.

Show where the number falls on the number line.

0 —————————————————— 100

Reflect

**SNAP for 0–1,000**

# SNAP
## Number Sense
### 0–1,000

Name:

Date:

Count forward by _____ from the number.

Show the value of the number.

Describe your picture.

Show the number in expanded form.

Create three equations that equal the number.

Create a real-life example that shows the value of the number.

Count backward by _____ from the number.

Show where the number falls on the number line.

0 — 1,000

Reflect

---

The SNAP Solution: An Innovative Math Assessment Tool for Grades K–8
Copyright © 2024 by Kirk Savage, Jonathan Ferris, and Tom Hierck
SolutionTree.com · Visit **go.SolutionTree.com/mathematics** to download this free reproducible.

## SNAP for 0–10,000

# SNAP
## Number Sense
### 0–10,000

Name: _____

Date: _____

**Show the value of the number.**

**Describe your picture.**

**Count backward by _____ from the number.**

**Show the number in expanded form.**

**Create three equations that equal the number.**

**Create a real-life example that shows the value of the number.**

**Count forward by _____ from the number.**

**Show where the number falls on the number line.**

0 — 10,000

**Reflect**

**SNAP for 0–100,000**

# SNAP
## Number Sense
**0–100,000**

Name:

Date:

- Show the value of the number.
- Describe your picture.
- Show the number in expanded form.
- Create three equations that equal the number.
- Create a real-life example that shows the value of the number.
- Count forward by _____ from the number.
- Count backward by _____ from the number.
- Show where the number falls on the number line. (0 to 100,000)
- Reflect

---

The SNAP Solution: An Innovative Math Assessment Tool for Grades K–8
Copyright © 2024 by Kirk Savage, Jonathan Ferris, and Tom Hierck
SolutionTree.com · Visit **go.SolutionTree.com/mathematics** to download this free reproducible.

**Open-Ended SNAP**

# SNAP
## Number Sense

Name: _____

Date: _____

- Show the value of the number.
- Describe your picture.
- Show the number in expanded form.
- Create three equations that equal the number.
- Create a real-life example that shows the value of the number.
- Count backward by _____ from the number.
- Count forward by _____ from the number.
- Show where the number falls on the number line.

**Reflect**

REPRODUCIBLE | 125

**Blank SNAP**

# SNAP
## Number Sense

Name: _____

Date: _____

---

**The SNAP Solution: An Innovative Math Assessment Tool for Grades K–8**
Copyright © 2024 by Kirk Savage, Jonathan Ferris, and Tom Hierck
SolutionTree.com · Visit **go.SolutionTree.com/mathematics** to download this free reproducible.

# SNAP Number Sense Rubric

| Competency | 1—Emerging<br>*Student understanding and application of learning standards are not evident.* | 2—Developing<br>*The student demonstrates some understanding and application of number sense.* | 3—Applying<br>*The student demonstrates proficient understanding and application of number sense.* | 4—Extending<br>*The student demonstrates insightful understanding and application.* |
|---|---|---|---|---|
| Conceptual understanding<br>*Picture box* | | | • Picture clearly communicates and represents the value of the number<br>• Accurate | |
| *Describe picture* | | | • Accurately describes and elaborates on pictorial representation<br>• Clearly communicates | |
| Procedural fluency<br>*Expanded form* | | | • Accurately demonstrates the value of each digit | |
| *Counting forward and backward* | | | • Complete and accurate | |
| Strategic competence<br>*Real-life connection* | | | • Provides connection to a real-life example<br>• Demonstrates understanding of the number value | |
| *Three equations* | | | • Accurately uses grade-appropriate operations in all three equations | |
| Adaptive reasoning<br>*Number line* | | | • Correctly estimates placement of number on provided number line with benchmarks (justification) | |
| Productive disposition<br>*Reflection* | | | • Some insight on mathematical thinking evident<br>• Evidence of self-efficacy in response | |
| **Cut Scores:** | 0–7: Emerging | 8–12: Developing | 13–16: Applying | 17–20: Extending |

REPRODUCIBLE | 127

# SNAP Number Sense Class Profile

Teacher: _____ Year: _____ Doorway question: _____

| Conceptual understanding | Procedural fluency | Strategic competence | Adaptive reasoning | Productive disposition |
|---|---|---|---|---|
| *Picture box and description picture* | *Expanded form and counting forward and backward* | *Real-life connection and three equations* | *Number line* | *Reflection* |
| Not evident    Proficient | Not evident    Proficient | Not evident    Proficient | Not evident    Proficient | Not evident    Proficient |

**Whole-class instruction**

**Small-group instruction**

**Individual instruction**

*The SNAP Solution: An Innovative Math Assessment Tool for Grades K–8*
Copyright © 2024 by Kirk Savage, Jonathan Ferris, and Tom Hierck
SolutionTree.com · Visit **go.SolutionTree.com/mathematics** to download this free reproducible.

# APPENDIX B

# Resources to Support Number Sense

After students complete the SNAP, knowing where to start and how to support students can overwhelm teachers. We recommend you start slowly as you begin your journey using this assessment tool. Throughout this book, we reference a variety of resources with a direct connection to the SNAP you will find valuable for classroom use. The following list of books and websites provides you with proven ways to assist students in the development of number sense.

## Books

### CHORAL COUNTING & COUNTING COLLECTIONS: TRANSFORMING THE PREK–5 MATH CLASSROOM

By Megan L. Franke, Elham Kazemi, and Angela Chan Turrou (2018)

This book provides teachers with valuable strategies for understanding mathematical concepts through counting. We recommend this book because it allows students to engage with numbers in playful and intentional ways that focus on increasing students' number sense.

### CLASSROOM-READY NUMBER TALKS FOR 6TH, 7TH, AND 8TH GRADE TEACHERS

By Nancy Hughes (2020)

This resource provides a series of strategies that aims to teach middle school students mental mathematics and problem-solving skills. We recommend this book because it continues to expose students to number talks, a strategy they first experience in the earlier grades.

### DAILY ROUTINES TO JUMP-START MATH CLASS, MIDDLE SCHOOL: ENGAGE STUDENTS, IMPROVE NUMBER SENSE, AND PRACTICE REASONING

By John J. SanGiovanni and Eric Milou (2018)

This book offers strategies to promote reasoning and number sense in the middle school classroom. We recommend this resource because its aim is to assist students in deepening their number sense in the late-intermediate grades.

### ELEMENTARY AND MIDDLE SCHOOL MATHEMATICS: TEACHING DEVELOPMENTALLY

By John A. Van de Walle, Karen S. Karp, and Jennifer M. Bay-Williams (2023)

This book not only outlines how students learn mathematics but also provides teachers many ways to teach mathematics through hands-on activities, inquiry, and problem solving. It also demonstrates how teachers can foster a positive mathematical environment in their classrooms. This book connects with Common Core State Standards (NGA & CCSSO, 2010a, 2010b) and the NCTM's *Principles to Actions: Ensuring Mathematical Success for All* (Leinwand, Brahier, & Huinker, 2024). We recommend all teachers have this resource in their classroom library.

### HIGH-YIELD ROUTINES FOR GRADES K–8

By Ann C. McCoy, Joann Barnett, and Emily Combs (2013)

This book demonstrates seven different mathematical routines teachers may use from kindergarten through grade 8. Each routine connects to mathematical content and competencies. This is an excellent resource to strengthen students' number sense.

### MATHEMATICAL MINDSETS: UNLEASHING STUDENTS' POTENTIAL THROUGH CREATIVE MATH, INSPIRING MESSAGES AND INNOVATIVE TEACHING

By Jo Boaler (with a foreword by Carol S. Dweck; 2016)

This book is groundbreaking in addressing the mathematics anxiety students often experience, and it provides ways for students to experience mathematics success. We highly recommend this book for its theoretical underpinnings and practical applications about mathematical thinking.

### NUMBER SENSE ROUTINES: BUILDING NUMERICAL LITERACY EVERY DAY IN GRADES K–3

By Jessica Shumway (2011)

This book contains a series of valuable five-, ten-, and fifteen-minute exercises to promote mathematical thinking. These daily warm-ups support young students in developing the foundations of number sense. Included are conversations among student peers that demonstrate the importance of listening to students and knowing what to look for. We recommend this book because it engages early learners in all facets of number sense (including how to subitize, estimate, and compute), plus it provides reasoning strategies and visual models to solve problems.

### NUMBER SENSE ROUTINES: BUILDING MATHEMATICAL UNDERSTANDING EVERY DAY IN GRADES 3–5

By Jessica Shumway (2018)

This book provides examples of implementing number sense routines in grades 3–5 and, as in the book covering the earlier grades, a series of five-, ten-, or fifteen-minute exercises to support students' number sense. We recommend this resource for use in the upper-elementary grades.

### NUMBER TALKS: HELPING CHILDREN BUILD MENTAL MATH AND COMPUTATION STRATEGIES, GRADES K–5

By Sherry Parrish (2010)

This book explains the value of having five-to-fifteen-minute guided classroom discussions about mathematics. We recommend this book because it promotes a growth mindset in mathematics and allows students to share their thinking in a risk-free environment.

### PLACE VALUE IN INTERMEDIATE: BUILDING NUMBER SENSE IN GRADES 3 TO 5

By Carole Fullerton (2017)

This resource is for grades 3–5 teachers with a focus on engaging students in deep learning regarding number sense. This book provides a series of open-ended tasks directed at having students represent and describe numbers, and compare and order numbers to 100,000, including explorations of decimal numbers to thousandths. We recommend this book because it promotes number sense in the intermediate grades.

### PLACE VALUE IN PRIMARY: DEVELOPING NUMBER SENSE IN KINDERGARTEN THROUGH GRADE 2

By Carole Fullerton (2016)

This book is for teachers of kindergarten through grade 2, including combined grades. The content includes lessons on quantity, comparing, ordering sets, estimating, and skip counting. It outlines ideas for whole-class and small-group instruction, including practice games that allow students to experience mathematics concepts in engaging ways. We recommend this book to early primary teachers who want to reinforce number sense concepts in engaging ways.

### THE *GOOD QUESTIONS* SERIES

By Carole Fullerton

These books provide mathematical entry points by posing questions to students in kindergarten through grade 8. The series addresses the key aspects of number sense with respect to subitizing, counting, and estimating, and provides open-ended tasks at the end of each learning sequence to inspire thinking and connections. We recommend these books to teachers as they build students' mathematical capacity by promoting inquiry and classroom discussion.

## Websites

### CLOTHESLINE MATH

(**https://clotheslinemath.com**): The *clothesline* is a manipulatable number line that makes facilitating class discourse on number sense much more efficient and effective. The clothesline is dynamic, so teachers may need to adjust the benchmark numbers, as well as the values on the line.

### DAILY ROUTINES AT TEACH AT THE SPEED OF LEARNING

(**https://visiblethinking.weebly.com/daily-routines.html**): Find instructional routines such as number talks, estimation tasks, dot cards, "Which One Doesn't Belong?", numberless word problems, and evergreen games.

### ESTIMATION 180 K–12

(**https://estimation180.com**): Daily estimation challenges help students improve both number sense and problem-solving skills.

### GFLETCHER

(**https://gfletchy.com**): Graham Fletcher, a mathematics specialist, is the creator of the site's fascinating and informative mathematics progression videos and three-act tasks.

### MATHEMATICAL THINKING BY CAROLE FULLERTON

(**https://mindfull.wordpress.com.**): Find games and resources to support student number sense development.

### NUMBER TALK IMAGES

(**http://ntimages.weebly.com**): This collaborative project gathers interesting real-life photographic images for teachers to use as a launching point for number talks.

### THE RECOVERING TRADITIONALIST

(**www.therecoveringtraditionalist.com**): This site has easy-to-follow tutorials and professional development videos from former middle school mathematics teacher Christina Tondevold, who helps other teachers learn what she knows about how students think about mathematics. Tondevold focuses on students' conceptual rather than procedural understanding.

### SAVAGEBIRDLEARNING

**(https://savagebirdlearning.com):** Kevin Bird and Kirk Savage created learning materials to assist teachers in effectively using performance standards–based assessments in K–12 education systems.

### SNAP 33

**(Visit https://snap.sd33.bc.ca):** The Chilliwack School District created this site to support teachers in implementing of the SNAP.

### STEVE WYBORNEY'S BLOG

**(https://stevewyborney.com):** Wyborney is well known for his use of instructional technology, mathematics work, and belief in the potential of every student.

### WHICH ONE DOESN'T BELONG?

**(https://wodb.ca):** This site features thought-provoking open-ended puzzles for mathematics students and teachers. Although the site does not provide answers, there are many ways to describe which one doesn't belong.

### YOUCUBED K–12

**(www.youcubed.org):** This site features brain science, mathematics mindset, visual tasks, and number talks for K–12 students.

# References and Resources

Alajmi, A., & Reys, R. (2007). Reasonable and reasonableness of answers: Kuwaiti middle school teachers' perspectives. *Educational Studies in Mathematics, 65*(1), 77–94.

Andrews, P., & Sayers, J. (2015). Identifying opportunities for grade one students to acquire foundational number sense: Developing a framework for cross-cultural classroom analyses. *Early Childhood Education Journal, 43*(4), 257–267. https://doi.org/10.1007/s10643-014-0653

Anno, M. (1977). *Anno's counting book.* New York: Crowell.

Askew, M., Rhodes, V., Brown, M., Wiliam, D., & Johnson, D. (1997). *Effective teachers of numeracy: Report of a study carried out for the Teacher Training Agency.* London: King's College, University of London.

Aunola, K., Leskinen, E., Lerkkanen, M.-K., & Nurmi, J.-E. (2004). Developmental dynamics of math performance from preschool to grade 2. *Journal of Educational Psychology, 96*(4), 699–713. https://doi.org/10.1037/0022-0663.96.4.699

Baroody, A. J., Eiland, M. D., Purpura, D. J., & Reid, E. E. (2012). Fostering at-risk kindergarten children's number sense. *Cognition and Instruction, 30*(4), 435–470.

Baroody, A. J., Eiland, M., & Thompson, B. (2009). Fostering at-risk preschoolers' number sense. *Early Education and Development, 20*(1), 80–128. https://doi.org/10.1080/10409280802206619

Beilock, S. L., Gunderson, E. A., Ramirez, G., & Levine, S. C. (2010). Female teachers' math anxiety affects girls' math achievement. *Proceedings of the National Academy of Sciences, 107*(5), 1860–1863.

Beswick, K., Muir, T., & McIntosh, A. (2004). *Developing an instrument to assess the number sense of young children* [Conference presentation]. Australian Association for Research in Education (AARE) Annual Conference, Melbourne, Australia.

Bieda, K. N., & Staples, M. E. (2020). Justification as an equity practice. *Mathematics Teacher: Learning and Teaching PK–12, 113*(2), 102–108.

Bird, K., & Savage, K. (2014). *The ANIE: A math assessment tool that reveals learning and informs teaching.* Markham, Ontario, Canada: Pembroke.

Bird, K., & Savage, K. (2015). *P.L.A.N. for better learning: 4 simple steps for designing lessons that boost thinking and maximize learning.* Markham, Ontario, Canada: Pembroke.

Björn, P. M., Aunola, K., & Nurmi, J.-E. (2016). Primary school text comprehension predicts mathematical word problem-solving skills in secondary school. *Educational Psychology, 36*(2), 362–377.

Black, P., & Wiliam, D. (2009). Developing the theory of formative assessment. *Educational Assessment, Evaluation and Accountability, 21*(1), 5–31.

Boaler, J. (1998). Open and closed mathematics: Student experiences and understandings. *Journal for Research in Mathematics Education, 29*(1), 41–62.

Boaler, J. (2012, July 3). *Timed tests and the development of math anxiety.* Accessed at www.edweek.org/teaching-learning/opinion-timed-tests-and-the-development-of-math-anxiety/2012/07 on December 27, 2023.

Boaler, J. (2014). Research suggests that timed tests cause math anxiety. *Teaching Children Mathematics, 20*(8), 469–474.

Boaler, J. (2015, January 28). *Fluency without fear: Research evidence on the best ways to learn math facts.* Accessed at www.youcubed.org/evidence/fluency-without-fear on February 28, 2024.

Boaler, J. (2016). *Mathematical mindsets: Unleashing students' potential through creative math, inspiring messages and innovative teaching.* San Francisco: Jossey-Bass.

Boaler, J., Munson, J., & Williams, C. (2021). *Mindset mathematics: Visualizing and investigating big ideas, grade 1.* San Francisco: Jossey-Bass.

Bodily, J. (2012). *A classroom experiment: Implementing a math-talk environment in a university setting* [Master's thesis, Utah State University]. USU Digital Commons. Accessed at https://digitalcommons.usu.edu/gradreports/321 on May 23, 2024.

Booth, J. L., & Siegler, R. S. (2008). Numerical magnitude representations influence arithmetic learning. *Child Development, 79*(4), 1016–1031. https://doi.org/10.1111/j.1467-8624.2008.01173.x

Boushey, G., & Moser, J. (2014). *The daily 5: Fostering literacy independence in the elementary grades* (2nd ed.). Portland, ME: Stenhouse.

Bower, G. H., & Morrow, D. G. (1990). Mental models in narrative comprehension. *Science, 247*(4938), 44–48.

Bruner, J. S. (1966). *Toward a theory of instruction.* Cambridge, MA: Belknap Press of Harvard University.

Buffum, A., Mattos, M., & Malone, J. (2018). *Taking action: A handbook for RTI at Work™.* Bloomington, IN: Solution Tree Press.

Buffum, A., Mattos, M., Malone, J., Cruz, L. F., Dimich, N., & Schuhl, S. (in press). *Taking action: A handbook for RTI at Work™* (2nd ed.). Bloomington, IN: Solution Tree Press.

Buffum, A., Mattos M., & Weber C. (2009). *Pyramid response to intervention: RTI, professional learning communities, and how to respond when kids don't learn.* Bloomington, IN: Solution Tree Press.

Buffum, A., Mattos M., & Weber C. (2012). *Simplifying response to intervention: Four essential guiding principles*. Bloomington, IN: Solution Tree Press.

Burns, M. (2007). *About teaching mathematics: A K–8 resource* (3rd ed.). Sausalito, CA: Math Solutions.

Çakır, H., & Cengiz, Ö. (2016). The use of open ended versus closed ended questions in Turkish classrooms. *Open Journal of Modern Linguistics, 6*(2), 60–70.

Campbell, J. I. D. (Ed.). (2005). *Handbook of mathematical cognition*. New York: Psychology Press.

Cawley, J. F., Parmar, R. S., Lucas-Fusco, L. M., Kilian, J. D., & Foley, T. E. (2007). Place value and mathematics for students with mild disabilities: Data and suggested practices. *Learning Disabilities: A Contemporary Journal, 5*(1), 21–39.

Chappuis, J., & Stiggins, R. (2020). *Classroom assessment for student learning: Doing it right—Using it well* (3rd ed.). New York: Pearson.

Chilliwack School District. (2016). *SNAP: Student numeracy assessment & practice*. Accessed at https://snap.sd33.bc.ca on July 28, 2023.

Clarke, B., & Shinn, M. R. (2004). A preliminary investigation into the identification and development of early mathematics curriculum-based measurement. *School Psychology Review, 33*(2), 234–248.

Close, S. (2001). *Pathways to powerful performance: Effective ways of supporting educators to integrate the systematic teaching of cognitive and metacognitive strategies into their practice of teaching writing* [Unpublished master's thesis]. Royal Roads University.

Cross, C. T., Woods, T. A., & Schweingruber, H. (Eds.). (2009). *Mathematics learning in early childhood: Paths toward excellence and equity*. Washington, DC: National Academies Press.

Duncan, G. J., Dowsett, C. J., Claessens, A., Magnuson, K., Huston, A. C., Klebanov, P., et al. (2007). School readiness and later achievement. *Developmental Psychology, 43*(6), 1428–1446.

Dweck, C. S. (2015). The secret to raising smart kids. *Scientific American Mind, 23*(5), 10–17.

Dyson, N. I., Jordan, N. C., & Glutting, J. (2013). A number sense intervention for low-income kindergartners at risk for mathematics difficulties. *Journal of Learning Disabilities, 46*(2), 161–181.

Faulkner, V. N. (2009). The components of number sense: An instructional model for teachers. *TEACHING Exceptional Children, 41*(5), 24–30.

Finlayson, M. (2014). Addressing math anxiety in the classroom. *Improving Schools, 17*(1), 99–115. https://doi.org/10.1177/1365480214521457

Fisher, R. (2005). *Teaching children to think* (2nd ed.). Cheltenham, United Kingdom: Nelson Thornes.

Fosnot, C. T., & Dolk, M. (2001). *Young mathematicians at work: Constructing number sense, addition, and subtraction*. Portsmouth, NH: Heinemann.

Franke, M. L., Kazemi, E., & Turrou, A. C. (Eds.) (2018). *Choral counting & counting collections: Transforming the preK–5 math classroom*. Portland, NH: Stenhouse.

Fullan, M. (2007). *The new meaning of educational change* (4th ed.). New York: Teachers College Press.

Fullan, M. (2016). *The new meaning of educational change* (5th ed.). New York: Teachers College Press.

Fullerton, C. (n.d.). *More good questions: A year of open-ended math problems for grades 5–8*. Vancouver, British Columbia, Canada: Mind-Full Math Resources.

Fullerton, C. (2016). *Place value in primary: Developing number sense in kindergarten through grade 2*. Vancouver, British Columbia, Canada: Mind-Full Math Resources.

Fullerton, C. (2017). *Place value in intermediate: Developing number sense in grades 3 to 5*. Vancouver, British Columbia, Canada: Mind-Full Math Resources.

Fullerton, C. (2018a). *Good questions: A year of open-ended math problems for grades 2–4*. Vancouver, British Columbia, Canada: Mind-Full Math Resources.

Fullerton, C. (2018b). *More good questions: A year of open-ended math problems for grades 5–8*. Vancouver, British Columbia, Canada: Mind-Full Math Resources.

Fullerton, C. (2021). *More good questions: A year of open-ended math problems for grades 2–4*. Vancouver, British Columbia, Canada: Mind-Full Math Resources.

Fullerton, C. (2022). *Good questions for kindergarten and grade 1: Lessons for building number sense*. Vancouver, British Columbia, Canada: Mind-Full Math Resources.

Fuson, K. C., Kalchman, M., & Bransford, J. D. (2005). Mathematical understanding: An introduction. In M. S. Donovan & J. D. Bransford (Eds.), *How students learn: History, mathematics, and science in the classroom* (pp. 217–256). Washington, DC: National Academies Press.

Fuson, K. C., & Murata, A. (2007). Integrating NRC principles and the NCTM process standards to form a class learning path model that individualizes within whole-class activities. *National Council of Supervisors of Mathematics Journal of Mathematics Education Leadership, 10*(1), 72–91.

Gainsburg, J. (2008). Real-world connections in secondary mathematics teaching. *Journal of Mathematics Teacher Education, 11*(3), 199–219. https://doi.org/10.1007/s10857-007-9070-8

Geary, D. C., Hoard, M. K., Byrd-Craven, J., Nugent, L., & Numtee, C. (2007). Cognitive mechanisms underlying achievement deficits in children with mathematical learning disability. *Child Development, 78*(4), 1343–1359.

Gersten, R., Jordan, N. C., & Flojo, J. R. (2005). Early identification and interventions for students with mathematics difficulties. *Journal of Learning Disabilities, 38*(4), 293–304.

Godin, S. (2007). *The dip: A little book that teaches you when to quit (and when to stick)*. New York: Portfolio.

Goodreads. (n.d.). *Maya Angelou quotes*. Accessed at https://goodreads.com/quotes/9821-i-did-then-what-i-knew-how-to-do-now on March 12, 2024.

Hattie, J. (2009). *Visible learning: A synthesis of over 800 meta-analyses relating to achievement*. New York: Routledge.

Hattie, J. (2023). *Visible learning: The sequel—A synthesis of over 2,100 meta-analyses relating to achievement*. New York: Routledge.

Hierck, T., & Weber, C. (2023). *The road to success with MTSS: A ten-step process for schools*. Bloomington, IN: Solution Tree Press.

Howell, S., & Kemp, C. (2005). Defining early number sense: A participatory Australian study. *Educational Psychology, 25*(5), 555–571. https://doi.org/10.1080/01443410500046838

Howell, S., & Kemp, C. (2009). A participatory approach to the identification of measures of number sense in children prior to school entry. *International Journal of Early Years Education, 17*(1), 47–65. https://doi.org/10.1080/09669760802699902

Howell, S. C., & Kemp, C. R. (2010). Assessing preschool number sense: Skills demonstrated by children prior to school entry. *Educational Psychology, 30*(4), 411–429. https://doi.org/10.1080/01443411003695410

Hughes, A. F., & Adera, B. (2006). Education and day treatment opportunities in schools: Strategies that work. *Preventing School Failure, 51*(1), 26–30.

Hughes, N. (2020). *Classroom-ready number talks for sixth, seventh, and eighth grade teachers: 1,000 interactive math activities that promote conceptual understanding and computational fluency*. Berkeley, CA: Ulysses Press.

Ivrendi, A. (2011). Influence of self-regulation on the development of children's number sense. *Early Childhood Education Journal, 39*(4), 239–247. https://doi.org/10.1007/s10643-011-0462-0

Jameson, M. M., & Fusco, B. R. (2014). Math anxiety, math self-concept, and math self-efficacy in adult learners compared to traditional undergraduate students. *Adult Education Quarterly, 64*(4), 306–322.

Johnson, B. L. (2018). *Math teachers who don't like math: A narrative inquiry into the mathematics experience of early childhood and elementary educators who dislike mathematics* [Doctoral dissertation, Ball State University]. ProQuest Dissertations. Accessed at www.proquest.com/docview/2088427300 on May 23, 2024.

Jordan, N. C., Glutting, J., Dyson, N., Hassinger-Das, B., & Irwin, C. (2012). Building kindergartners' number sense: A randomized controlled study. *Journal of Educational Psychology, 104*(3), 647–660. https://doi.org/10.1037/a0029018

Jordan, N. C., Glutting, J., & Ramineni, C. (2010). The importance of number sense to mathematics achievement in first and third grades. *Learning and Individual Differences, 20*(2), 82–88. https://doi.org/10.1016/j.lindif.2009.07.004

Jordan, N. C., Glutting, J., Ramineni, C., & Watkins, M. W. (2010). Validating a number sense screening tool for use in kindergarten and first grade: Prediction of mathematics proficiency in third grade. *School Psychology Review, 39*(2), 181–195.

Jordan, N. C., Kaplan, D., Oláh, L. N., & Locuniak, M. N. (2006). Number sense growth in kindergarten: A longitudinal investigation of children at risk for mathematics difficulties. *Child Development, 77*(1), 153–175.

Keene, E. O., & Zimmerman, S. (2007). *Mosaic of thought: The power of comprehension strategy instruction* (2nd ed.). Portsmouth, NH: Heinemann.

Khoshaim, H. B. (2020). Mathematics teaching using word-problems: Is it a phobia! *International Journal of Instruction, 13*(1), 855–868. https://doi.org/10.29333/iji.2020.13155a

Landerl, K., Bevan, A., & Butterworth, B. (2004). Developmental dyscalculia and basic numerical capacities: A study of 8-9-year-old students. *Cognition, 93*(2), 99–125.

Lee, Y., Kinzie, M. B., & Whittaker, J. V. (2012). Impact of online support for teachers' open-ended questioning in pre-K science activities. *Teaching and Teacher Education, 28*(4), 568–577.

Leinwand, S., Brahier, D. J., & Huinker, D. (2024). *Principles to actions: Ensuring mathematical success for all* (10th anniversary ed.). Reston, VA: National Council of Teachers of Mathematics.

Lesh, R. (1979). Mathematical learning disabilities: Considerations for identification, diagnosis, and remediation. In R. Lesh, D. Mierkiewicz, & M. Kantowski (Eds.), *Applied mathematical problem solving* (pp. 111–180). Columbus, OH: ERIC/SMEAC.

Louange, J., & Bana, J. (2010, July). *The relationship between the number sense and problem solving abilities of year 7 students*. Accessed at https://files.eric.ed.gov/fulltext/ED520910.pdf on March 8, 2024.

Maghfirah, M., & Mahmudi, A. (2018). Number sense: The result of mathematical experience. *Journal of Physics: Conference Series, 1097*(1). https://doi.org/10.1088/1742-6596/1097/1/012141

Malofeeva, E., Day, J., Saco, X., Young, L., & Ciancio, D. (2004). Construction and evaluation of a number sense test with Head Start children. *Journal of Educational Psychology, 96*(4), 648–659. https://doi.org/10.1037/0022-0663.96.4.648

Martin, W. G., Carter, J., Forster, S., Howe, R., Kader, G., Kepner, H., et al. (2009). *Focus in high school mathematics: Reasoning and sense making*. Reston, VA: National Council of Teachers of Mathematics.

Marzano, R. J. (2003). *What works in schools: Translating research into action*. Alexandria, VA: ASCD.

McClure, N. (2021). *1, 2, 3 Salish Sea: A Pacific Northwest counting book*. Seattle, WA: Little Bigfoot.

McCoy, A. C., Barnett, J., & Combs, E. (2013). *High-yield routines for grades K–8*. Reston, VA: National Council of Teachers of Mathematics.

McGuire, P., Kinzie, M. B., & Berch, D. B. (2012). Developing number sense in pre–K with five-frames. *Early Childhood Education Journal*, *40*(4), 213–222.

McIntosh, A., Reys, B. J., & Reys, R. E. (1992). A proposed framework for examining basic number sense. *For the Learning of Mathematics*, *12*(3), 2–8.

Mensah, J. K., Okyere, M., & Kuranchie, A. (2013). Student attitude towards mathematics and performance: Does the teacher attitude matter? *Journal of Education and Practice*, *4*(3), 132–139.

Mills, M., & Stevens, P. (1998). *Improving writing and problem solving skills of middle school students* [Master's thesis, Saint Xavier University]. ERIC Institute of Education Sciences. Accessed at https://files.eric.ed.gov/fulltext/ED420876.pdf on May 23, 2024.

Moule, J. (2012). *Cultural competence: A primer for educators* (2nd ed.). Belmont, CA: Wadsworth.

National Council of Teachers of Mathematics. (2000). *Principles and standards for school mathematics*. Reston, VA: Author.

National Council of Teachers of Mathematics. (2014, April). *Algebra as a strand of school mathematics for all students: A position of the National Council of Teachers of Mathematics* [Position statement]. Accessed at www.nctm.org/uploadedFiles/Standards_and_Positions/Position_Statements/Algebra_2014-04.pdf on February 28, 2024.

National Governors Association Center for Best Practices & Council of Chief State School Officers. (2010a). *Common Core State Standards for English language arts and literacy in history/social studies, science, and technical subjects*. Washington, DC: Authors. Accessed at https://corestandards.org/wp-content/uploads/2023/09/ELA_Standards1.pdf on March 13, 2024.

National Governors Association Center for Best Practices & Council of Chief State School Officers. (2010b). *Common Core State Standards for mathematics*. Washington, DC: Authors. Accessed at https://corestandards.org/wp-content/uploads/2023/09/Math_Standards1.pdf on November 27, 2023.

National Research Council. (2001). *Adding it up: Helping children learn mathematics*. Washington, DC: National Academy Press.

Neergaard, L. (2013, March 25). *Early number sense plays role in later math skills*. Accessed at https://phys.org/news/2013-03-early-role-math-skills.html on November 13, 2023.

Newhouse, M. (2017). *Counting on snow* (Reissue). Toronto, Ontario, Canada: Tundra Books. (Original work published 2010)

Novakowski, J. (2007). Early childhood corner: Developing five-ness in kindergarten. *Teaching Children Mathematics*, *14*(4), 226–231.

O'Neil, H. F., Jr., & Brown, R. S. (1998). Differential effects of question formats in mathematics assessment on metacognition and affect. *Applied Measurement in Education*, *11*(4), 331–351.

Organisation for Economic Co-operation and Development. (2004). *Learning for tomorrow's world: First results from PISA 2003*. Paris: Author. Accessed at https://oecd.org/education/school/programmeforinternationalstudentassessmentpisa/34002216.pdf on March 8, 2024.

Organisation for Economic Co-operation and Development. (2012). *Equity and quality in education: Supporting disadvantaged students and schools*. Paris: Author.

Organisation for Economic Co-operation and Development. (2019). *PISA 2018 results (Volume I): What students know and can do*. Paris: Author. Accessed at https://doi.org/10.1787/5f07c754-en on November 13, 2023.

Özsoy, G. (2011). An investigation of the relationship between metacognition and mathematics achievement. *Asia Pacific Education Review, 12*(2), 227–235.

Özsoy, G., & Ataman, A. (2009). The effect of metacognitive strategy training on mathematical problem solving achievement. *International Electronic Journal of Elementary Education, 1*(2), 68–83.

Palmer, P. J. (2017). *The courage to teach: Exploring the inner landscape of a teacher's life* (20th anniversary ed.). Hoboken, NJ: Jossey-Bass.

Parrish, S. (2010). *Number talks: Helping children build mental math and computation strategies, grades K–5.* Sausalito, CA: Math Solutions.

Parrish, S. D. (2011). Number talks build numerical reasoning. *Teaching Children Mathematics, 18*(3), 198–206.

Pittalis, M., Pitta-Pantazi, D., & Christou, C. (2015, February). The development of student's early number sense. In K. Krainer & N. Vondrová (Eds.), *Proceedings of the ninth congress of the European Society for Research in Mathematics Education* (pp. 446–452). Prague, Czech Republic: Charles University.

Porter, B. E. (2019). *Elementary teachers' perceptions of teaching mathematics, mathematics anxiety, and teaching mathematics efficacy* [Doctoral dissertation, Marshall University]. Marshall Digital Scholar. Accessed at https://mds.marshall.edu/etd/1242 on May 21, 2024.

Pratt, S. S. (2018). Area models to image integer and binomial multiplication. *Investigations in Mathematics Learning, 10*(2), 85–105.

Putri, H. E., Suwangsih, E., Rahayu, P., Nikawanti, G., Enzelina, E., & Wahyudy, M. A. (2020). Influence of concrete-pictorial-abstract (CPA) approach on the enhancement of primary school students' mathematical reasoning ability. *Mimbar Sekolah Dasar (Elementary School Pulpit), 7*(1), 119–132. https://doi.org/10.17509/mimbar-sd.v7i1.22574

Reys, R. E., Reys, B. J., Nohda, N., & Emori, H. (1995). Mental computation performance and strategy use of Japanese students in grades 2, 4, 6, and 8. *Journal for Research in Mathematics Education, 26*(4), 304–326.

Reys, R. E., Rybolt, J. F., Bestgen, B. J., & Wyatt, J. W. (1982). Processes used by good computational estimators. *Journal for Research in Mathematics Education, 13*(3), 183–201.

Reys, R. E., & Yang, D.-C. (1998). Relationship between computational performance and number sense among sixth- and eighth-grade students in Taiwan. *Journal for Research in Mathematics Education, 29*(2), 225–237.

Roberts, N. (2019). The standard written algorithm for addition: Whether, when and how to teach it. *Pythagoras, 40*(1), a487.

Rohrer, D., Dedrick, R. F., & Stershic, S. (2015). Interleaved practice improves mathematics learning. *Journal of Educational Psychology, 107*(3), 900–908.

SanGiovanni, J. J., & Milou, E. (2018). *Daily routines to jump-start math class, middle school: Engage students, improve number sense, and practice reasoning.* Thousand Oaks, CA: Corwin.

Saundry, C., & Nicol, C. (2006, July 16–21). Drawing as problem solving: Young children's mathematical reasoning through pictures. In J. Novotná, H. Moraová, M. Krátká, & N. Stehlíková (Eds.), *Proceedings of the 30th conference of the International Group for the Psychology of Mathematics Education* (Vol. 5, pp. 57–63). Prague, Czech Republic: International Group for the Psychology of Mathematics Education.

Savage, K. (2021). *SNAP vs. FSA: Competency-based numeracy assessments count* [Doctoral dissertation, University of Kansas]. ProQuest Dissertations. Accessed at www.proquest.com/openview /392589ac29cb3b7423d678137434b676/1 on May 23, 2024.

Schneider, M., Merz, S., Stricker, J., De Smedt, B., Torbeyns, J., Verschaffel, L., et al. (2018). Associations of number line estimation with mathematical competence: A meta-analysis. *Child Development, 89*(5), 1467–1484. https://doi.org/10.1111/cdev.13068

Seely, C. L. (2009). *Faster isn't smarter: Messages about math, teaching and learning in the 21st century.* Sausalito, CA: Math Solutions.

Shanley, L., Clarke, B., Doabler, C. T., Kurtz-Nelson, E., & Fien, H. (2017). Early number skills gains and mathematics achievement: Intervening to establish successful early mathematics trajectories. *The Journal of Special Education, 51*(3), 177–188.

Sharma, S. (2015). Promoting risk taking in mathematics classrooms: The importance of creating a safe learning environment. *The Mathematics Enthusiast, 12*(1), 290–306.

Shore, K. (2015). *Dr. Ken Shore's classroom problem solver: Math anxiety.* Accessed at www.educationworld .com/a_curr/shore/shore066.shtml on February 26, 2024.

Shumway, J. F. (2011). *Number sense routines: Building numerical literacy every day in grades K–3.* Portland, ME: Stenhouse.

Shumway, J. F. (2018). *Number sense routines: Building mathematical understanding every day in grades 3–5.* Portland, ME: Stenhouse.

Siegler, R. S., & Booth, J. L. (2005). Development of numerical estimation: A review. In J. I. D. Campbell (Ed.), *Handbook of mathematical cognition* (pp. 197–212). New York: Psychology Press.

Simms, V., Clayton, S., Cragg, L., Gilmore, C., & Johnson, S. (2016). Explaining the relationship between number line estimation and mathematical achievement: The role of visuomotor integration and visuospatial skills. *Journal of Experimental Child Psychology, 145*, 22–33.

Simonsen, B., Fairbanks, S., Briesch, A., Myers, D., & Sugai, G. (2008). Evidence-based practices in classroom management: Considerations for research to practice. *Education and Treatment of Children, 31*(3), 351–380.

Singleton, G. E. (2015). *Courageous conversations about race: A field guide for achieving equity in schools* (2nd ed.). Thousand Oaks, CA: Corwin.

Singleton, G. E. (2022). *Courageous conversations about race: A field guide for achieving equity in schools and beyond* (3rd ed.). Thousand Oaks, CA: Corwin.

Small, M. (2012). *Making math meaningful to Canadian students, K–8.* Toronto, Ontario, Canada: Nelson Education.

Sowder, J. T. (1992). Making sense of numbers in school mathematics. In G. Leinhardt, R. Putman, & R. A. Hattrup (Eds.), *Analysis of arithmetic for mathematics teaching* (pp. 1–51). Hillsdale, NJ: Erlbaum.

Stiggins, R. J. (2002). Assessment crisis: The absence of assessment *for* learning. *Phi Delta Kappan, 83*(10), 758–765.

Stiggins, R. J., Arter, J. A., Chappuis, J., & Chappuis, S. (2004). *Classroom assessment for student learning: Doing it right—Using it well.* Portland, OR: Assessment Training Institute.

Stoehr, K. J., Turner, E., & Sugimoto, A. T. (2015, November 5–8). *One teacher's understandings and practices for making real-world connections in mathematics* [Conference presentation]. In T. G. Bartell, K. N. Bieda, R. T. Putnam, K. Bradfield, & H. Dominguez (Eds.), *Proceedings of the 37th annual meeting of the North American chapter of the International Group for the Psychology of Mathematics Education* (pp. 1150–1153). East Lansing, MI.

Su, H. F. H., Ricci, F. A., & Mnatsakanian, M. (2016). Mathematical teaching strategies: Pathways to critical thinking and metacognition. *International Journal of Research in Education and Science, 2*(1), 190–200.

Sugimoto, A. T., Turner, E. E., & Stoehr, K. J. (2017). A case study of dilemmas encountered when connecting middle school mathematics instruction to relevant real world examples. *Middle Grades Research Journal, 11*(2), 61–82.

Thomas, K. R. (2006). Students THINK: A framework for improving problem solving. *Teaching Children Mathematics, 13*(2), 86–95.

Uttal, D. H., Scudder, K. V., & DeLoache, J. S. (1997). Manipulatives as symbols: A new perspective on the use of concrete objects to teach mathematics. *Journal of Applied Developmental Psychology, 18*(1), 37–54.

Van de Walle, J. A., Karp, K. S., & Bay-Williams, J. M. (2013). *Elementary and middle school mathematics: Teaching developmentally* (8th ed.). Boston: Pearson.

Van de Walle, J. A., Karp, K. S., & Bay-Williams, J. M. (2023). *Elementary and middle school mathematics: Teaching developmentally* (11th ed.). Boston: Pearson.

Vilenius-Tuohimaa, P. M., Aunola, K., & Nurmi, J.-E. (2008). The association between mathematical word problems and reading comprehension. *Educational Psychology, 28*(4), 409–426. https://doi.org/10.1080/01443410701708228

Vlassis, J., Baye, A., Auquière, A., de Chambrier, A.-F., Dierendonck, C., Giauque, N., et al. (2023). Developing arithmetic skills in kindergarten through a game-based approach: A major issue for learners and a challenge for teachers. *International Journal of Early Years Education, 31*(2), 419–434.

Waite, L. A. (2013). *Re-awakening wonder: Creativity in elementary mathematics* [Doctoral dissertation, University of Lethbridge]. University of Lethbridge Archives. Accessed at https://hdl.handle.net/10133/3344 on May 24, 2024.

Whitacre, I., Henning, B., & Atabaş, Ş. (2020). Disentangling the research literature on number sense: Three constructs, one name. *Review of Educational Research, 90*(1), 95–134.

Wiliam, D. (2018). *Embedded formative assessment* (2nd ed.). Bloomington, IN: Solution Tree Press.

Woods, D. A., Ketterlin Geller, L., & Basaraba, D. (2017). Number sense on the number line. *Intervention in School and Clinic, 53*(4), 229–236.

Yilmaz, Z. (2017). Young children's number sense development: Age related complexity across cases of three children. *International Electronic Journal of Elementary Education, 9*(4), 891–902.

Yun, C., Havard, A., Farran, D. C., Lipsey, M. W., Bilbrey, C., & Hofer, K. G. (2011, July 20–23). Subitizing and mathematics performance in early childhood. In L. Carlson, T. F. Shipley, & C. Hoelscher (Eds.), *Proceedings of the 33rd annual meeting of the Cognitive Science Society* (pp. 680–684). Austin, TX: Cognitive Science Society.

Zevenbergen, R., Niesche, R., Grootenboer, P., & Boaler, J. (2008, June 28–July 1). *Creating equitable practice in diverse classrooms: Developing a tool to evaluate pedagogy.* In M. Goos, R. Brown, & K. Makar (Eds.), *Navigating currents and charting directions: Proceedings of the 31st annual conference of the Mathematics Education Research Group of Australasia* (pp. 637–643). Brisbane, Queensland, Australia: Mathematics Education Research Group of Australia.

# Index

**A**

*About Teaching Mathematics* (Burns), 30

adaptive reasoning, 19, 41, 55–57

*Adding It Up: Helping Children Learn Mathematics* (National Research Council), 19

Angelou, M., 22

ANIE, The, 7, 15, 16

arrays, 45

artistic and themed SNAP templates, 101, 102–104

assessments.
 *See also* SNAP (student numeracy assessment and practice)
   assessment data and interventions, 97
   common assessments, 97, 109
   data analysis and monitoring, 111–112
   formative assessments, 25
   open versus closed assessments, 20–21
   summative assessments, 26
   traditional assessment methods, moving away from, 3–4

automaticity, 46

**B**

Bana, J., 115

Basaraba, D., 55–56

base ten, 47

Bieda, K., 45–46

Boaler, J., 2, 9–10, 20

Bodily, J., 55

Buffum, A., 95, 96, 97

Burns, M., 30

**C**

Cawley, J., 47

Christou, C., 53

class profile example of a mixed number task
   about, 73
   recommendations for, 81–83

student A, 74–76
student B, 77–79
student C, 80–81, 82
class profile example of a whole-number task
   about, 83
   recommendations for, 92–94
   student 1, 84–86
   student 2, 87–89
   student 3, 90–92
class profile templates, 73, 99
closed assessments, 5, 20–21. *See also* assessments
clothesline mathematics activities, 56, 132
common assessments, 97, 109. *See also* assessments
competency assessments, 5. *See also* assessments
conceptual understanding
   about, 43
   drawing to represent numbers, 43–45
   strands of mathematical proficiency, 19, 41
   writing to describe your picture, 45–46
concrete, pictorial, and abstract approaches, 20
connectionist approaches, 3
counting
   counting collections, 69
   counting forward and backward from the number, 48–50, 68–69
   number sense skills, 12
   skip counting, 17, 48–50
cultural competence, 22–23
cut scores, 112

### D

data analysis and monitoring, 111–112
decomposing the number, 31, 32, 47, 64, 67–68
design process and the SNAP, 15

discovery approaches, 3
doorway questions, 21
drawing, 17, 43–45, 67
Dweck, C., 100

### E

Emori, H., 39–40
English language arts (ELA), 14, 19–20
equal sharing, 45
equations, 17
error hunts, 100
estimation, 13
expanded form of the number, 47–48
exploring rubrics, assessment, and competency-based learning
   about, 71
   chapter summary, 94
   class profile example of a mixed number task, 73–83
   class profile example of a whole-number task, 83–94
   understanding the SNAP rubric, 71–73
exploring the SNAP. *See also* SNAP (student numeracy assessment and practice)
   about, 9–10
   chapter summary, 27
   classroom applications, 23, 25–27
   features of the SNAP, 19–23
   strands of mathematical proficiency, 19
   why the SNAP works, 10, 12–13, 15, 17–18

### F

faking skills, 26–27
five and ten, building the idea of, 63–64, 65, 67
five strands of mathematical proficiency. *See* strands of mathematical proficiency
fluency, 9, 19–20. *See also* number sense; procedural fluency
Foley, T., 47

formative assessments, 25.
    *See also* assessments
Friends of Ten, 65
Fullan, M., 110
Fusco, B., 2

**G**

Glutting, J., 116
grouping students, 105.
    *See also* small-group instruction
growth mindset, 100
guess my number, 105

**H**

Hierck, T., 96
hundreds chart, 50

**I**

imaging strategies, 20
implementing the SNAP with
    beginning mathematics.
    *See also* SNAP (student numeracy
    assessment and practice)
    about, 61
    chapter summary, 70
    implementing in kindergarten
        and grade 1, 61–62
    using the SNAP in grade 1, 66–69
    using the SNAP in kindergarten, 63–65
individual instruction, class profile examples,
    82–83, 94
individualized education plans (IEPs), 27
interventions, 13, 97.
    *See also* response to intervention (RTI)
introduction
    about this book, 7
    mathematics anxiety, 1–2
    moving toward the SNAP, 5
    traditional assessment methods, 3–4

**J**

Jameson, M., 2
Johnson, B., 1
Jordan, N., 116

justifying thinking, 45–46, 69

**K**

Ketterlin Geller, L., 55–56
Kilian, J., 47

**L**

learning progressions, 17
Lesh, R., 15
lesson design and the SNAP, 18
looking at number sense foundations.
    *See* number sense
looking at school and district
    implementation.
    *See* school and district implementation
loose parts, 67
Louange, J., 115
low floor, high ceiling tasks, 20
Lucas-Fusco, L., 47

**M**

mad minutes, 3
Maghfirah, M., 51
Mahmudi, A., 51
Malone, J., 96, 97
manipulatives, 20
mathematics anxiety, 1–2, 23, 25
mathematics talks, 23
Mattos, M., 95, 96, 97
McIntosh, A., 30–31
measurement, 52
mental models, 13, 20
metacognition, 19, 22, 58
mindsets, 100
mix and mingle, 105
Mnatsakanian, M., 58
morning meeting strategy, 68–69
Moule, J., 22
multiplication area models, 45
multirepresentational approaches, 20
multitiered system of supports (MTSS), 96
Munson, J., 20

**N**

National Research Council, 19, 31, 43, 46–47, 51, 55, 57

Neergaard, L., 33

Nicol, C., 43–44

Nohda, N., 39–40

non-examples, 65

Novakowski, J., 63

number comparisons, 13

number lines
    adaptive reasoning and, 55–56
    circling the number on, 65
    content and item descriptors, 17
    number sense skills and, 12
    showing the quantity on the number line, 69
    skip counting and, 49
    SNAP and, 10, 25
    values on a number line, 57

number operations, 13

number patterns, 12

number recognition, 12

number sense
    about, 29–30
    chapter summary, 40
    classrooms and, 40
    content and descriptors, 17
    definition of, 9, 30–31
    example number sense and operation skills in grades 2-7, 110–111
    in grades 1-5, 33–36
    in grades 6-8, 36–40
    in kindergarten, 31–33
    moving toward the SNAP, 5
    number sense skills, 12–13, 15
    number sense SNAP rubric, 72
    practitioner's perspective, 33, 36, 39
    resources to support, 129–133
    SNAP number sense class profile template, 73
    theoretical perspective on, 30–31

number talks, 23, 44

number walks, 52

**O**

one more or less, building the idea of, 63–64, 67

open assessments, 5, 20–21. *See also* assessments

open-ended questions, 21

Organisation for Economic Co-operation and Development (OECD), 109

**P**

Palmer, P., 2–3

Parmar, R., 47

Parrish, S., 44

part-part-whole, 31, 47, 64, 67

Pittalis, M., 53

Pitta-Pantazi, D., 53

pop quizzes, 3

procedural fluency
    about, 46–47
    counting forward and backward from the number, 48–50
    strands of mathematical proficiency, 19, 41
    writing the number in expanded form, 47–48

productive dispositions, 19, 41, 57–59

*Pyramid Response to Intervention* (Buffum, Mattos, and Weber), 95

**Q**

questions
    doorway questions, 21
    open-ended questions, 21

**R**

Ramineni, C., 116

reading fluency, 19–20

real-life connections, 21

real-life examples
    content and item descriptors, 17

talking about examples that shows the number, 68
talking about mathematics in the world, 65
writing real-life examples of the number, 51–53
reflections, 17, 58
rekenreks, 31, 54, 57, 67, 68
relative size, 45
representing or subitizing the number, 64
reproducibles for
  blank SNAP, 125
  grade 1 number sense SNAP, 119
  kindergarten number sense SNAP, 118
  open-ended SNAP, 124
  SNAP for 0-100, 120
  SNAP for 0-1,000, 121
  SNAP for 0-10,000, 122
  SNAP for 0-100,000, 123
  SNAP number sense class profile, 127
  SNAP number sense rubric, 126
response to intervention (RTI)
  about, 95–96
  chapter summary, 108
  RTI pyramid of, 96
  RTI tiers and the SNAP connections, 98
  supporting RTI, 96–98
  supporting Tier 1, 99–101, 105
  supporting Tier 2, 105–106
  supporting Tier 3, 108
Reys, B., 30–31, 39–40
Reys, R., 30–31, 39–40
Ricci, F., 58
Roberts, N., 47–48
rubrics, understanding the SNAP rubric, 71–73. See also exploring rubrics, assessment, and competency-based learning

## S

Saundry, C., 43–44

school and district implementation. See also SNAP (student numeracy assessment and practice)
  about, 109–110
  chapter summary, 114
  cut scores, 112
  data analysis and monitoring, 111–112
  implementation at the school or district level, 110–111
  school leadership, 113–114
  training, 113
  transparency and accountability, 112–113
scores, recommended cut scores, 26
self-efficacy, 57
Sharma, S., 40
Singleton, G., 22–23
skip counting, 17, 48–50. See also counting
small-group instruction
  class profile examples for, 81–82, 94
  supporting RTI and, 97, 106, 108
  using the SNAP for, 26
SNAP (student numeracy assessment and practice). See also exploring the SNAP; implementing the SNAP with beginning mathematics; school and district implementation
  artistic and themed SNAP templates, 101, 102–104
  class profile sheet for the SNAP template, 99
  example of, 6
  example of for class profiles, 74, 77, 80
  example of for grade 1, 66
  example of for grade 4, 11, 24
  example of for grade 6, 38
  example of for grades 2-5, 35
  example of for kindergarten, 32, 63
  example of in the middle grades, 42
  example of with highlighting areas of writing, 14
  moving toward, 5

SNAP basis in research on assessments, 17–19

traditional assessment methods and, 4

understanding the SNAP rubric, 71–73

use of in grades 1-5, 34, 66–69

use of in grades 6-8, 36–37

use of in kindergarten, 63–65

SNAP fatigue, 34

Sower, J., 30

Staples, M., 45–46

Stiggins, R., 3

strands of mathematical proficiency

about, 41

adaptive reasoning, 55–57

chapter summary, 59

conceptual understanding, 43–46

five strands of, 19

procedural fluency, 46–50

productive dispositions, 57–59

strategic competence, 51–54

strategic competence

about, 51

creating three equations that equal the number, 53–54

strands of mathematical proficiency, 19, 41

writing real-life examples of the number, 51–53

Su, H., 58

subitizing, 64

summative assessments, 26. See also assessments

survivor students, 26–27

## T

*Taking Action: A Handbook for RTI at Work* (Buffum, Mattos, and Malone), 96

tallying, 67

teachers as facilitators, 34

ten frames, 63, 68

Thomas, K., 22

tiers of intervention. See response to intervention (RTI)

timed drills, 3

traditional assessment methods, 3–4. See also assessments

transmission approaches, 3

## U

understanding how the SNAP supports response to intervention. See response to intervention (RTI)

understanding the five strands of mathematical proficiency. See strands of mathematical proficiency

unit tests, 3

## W

Weber, C., 95

whole-class instruction

class profile examples for, 81, 92–94

supporting Tier 1, 100

using the SNAP as a summative assessment, 26

Wiliam, D., 4

Williams, C., 20

Woods, D., 55–56

word problems, 3–4, 13, 100

## Y

Yilmaz, Z., 29–30

Yun, C., 64

## Z

zooming in, 25, 100, 107

### Nurturing Math Curiosity With Learners in Grades K–2
*Chepina Rumsey and Jody Guarino*
Gain the educational tools needed for planning, communicating, and representing mathematical ideas to students. This book gives teachers instructional strategies to enhance their students' natural wonder and curiosity toward math concepts.
**BKG180**

### See It, Say It, Symbolize It
*Patrick L. Sullivan*
Anyone can learn mathematics and stay in the math game for life once they learn key superpowers that can demystify foundational concepts—from whole numbers, fractions, and place value operations to ratios, proportions, and percentages. This book offers teaching methods to develop a dynamic and flexible understanding of numbers and operations in young learners.
**BKG187**

### Common Denominators
*Jennifer A. Lenhardt*
"*Common Denominators* is a collection of stories braided together with research-informed strategies and tools," writes author Jennifer A. Lenhardt. Make sense of student engagement and belonging by using mathematics concepts that illustrate our common humanity and illuminate a clear, sustainable path for honoring and meeting all students' needs.
**BKG179**

### The Fact Tactics™ Fluency Program
*Juli K. Dixon*
Support students in developing a deep understanding of multiplication by emphasizing procedural fluency to develop automaticity in mathematics. This book will lead you through a 20-week program that utilizes six tactics to promote reasoning over rote memorization.
**BKG125**

### You're a Teacher Now! What's Next?
*Tom Hierck and Alex Kajitani*
Anyone who has ever held the noble title of teacher has had that initial moment where they ask themselves, "What have I got myself into?", write Tom Hierck and Alex Kajitani. This is your lifeline: a quick read and conversational resource that both new and veteran educators will find handy to have in the classroom.
**BKG142**

## Solution Tree | Press

a division of Solution Tree

Visit SolutionTree.com or call 800.733.6786 to order.

# Global PD teams
### Collaborative Learning for School Improvement

# Quality team learning **from authors you trust**

---

Global PD Teams is the first-ever **online professional development resource designed to support your entire faculty on your learning journey.** This convenient tool offers daily access to videos, mini-courses, eBooks, articles, and more packed with insights and research-backed strategies you can use immediately.

**GET STARTED**
SolutionTree.com/**GlobalPDTeams**
800.733.6786